Translating Nature Terminology

Translating Nature Terminology

By

Wojciech Kasprzak

CAMBRIDGE SCHOLARS
PUBLISHING

Translating Nature Terminology, by Wojciech Kasprzak

This book first published 2011

Cambridge Scholars Publishing

12 Back Chapman Street, Newcastle upon Tyne, NE6 2XX, UK

British Library Cataloguing in Publication Data
A catalogue record for this book is available from the British Library

ISBN (10): 1-4438-2850-5, ISBN (13): 978-1-4438-2850-5

TABLE OF CONTENTS

List of Tables

ACKNOWLEDGEMENTS

I would like to thank Professor Aniela Korzeniowska for starting me off on this project, reading the manuscript in parts and as a whole and making helpful comments and corrections.

I would also like to express my gratitude to Professor Elżbieta Tabakowska, Professor Krzysztof Hejwowski and Professor Jan Rusiecki, whose critical observations have helped this work to take its ultimate form. The responsibility for any remaining shortcomings is entirely mine.

My thanks also go to Dr Zbigniew Możejko for his kind advice and encouragement.

I acknowledge the forbearance shown by my wife, Sylwia, during the many long hours I spent at the computer and the role played by my children, Staszka and Zdziś, as providers of some much needed healthy distraction from my work.

NOTES TO THE READER

1. The following abbreviations are used for the sake of convenience:

Publications

OALD	Cowie, A P, ed. 1989. *Oxford advanced learner's dictionary*. Oxford: OUP.
CED	Sinclair, John, ed. 1995. *Collins Cobuild English dictionary*. London: HarperCollins Publishers.
Webster's	Gove, Babcock Philip, ed. 1993. *Webster's third new international dictionary*. Cologne: Konemann.
BBC	Drive Publications Limited, eds. 1981. *Book of the British countryside. Great Britain*: DPL.
BM	Blamey, Marjorie, Richard Fitter, and Alastair Fitter. 2003. *Wild flowers of Britain and Ireland*. London: A & C Black.
JAK	Anioł-Kwiatkowska, Jadwiga. 2003. *Wielojęzyczny słownik florystyczny*. Wrocław: Wyd. Uniwersytetu Wrocławskiego.
KCz	Czekierda, Krzysztof. 1992. *Słownik ekologiczny polsko-angielski*. Warszawa: Terra Sana. Czekierda, Krzysztof. 1995. *Słownik ekologiczny angielsko-polski*. Białystok: Wyd. Ekonomia i Środowisko.

Other than publications

BAP	Biodiversity Action Plan
CORINE	Coordination of Information on the Environment; the international programme owing to which the PHYSIS (not an acronym, to my knowledge) database of natural habitats has been created.
EUNIS	European University Information Systems (used below to identify one of the databases of natural habitats)
NBN	National Biodiversity Network
NVC	National Vegetation Classification

2. The analysed sources are often bilingual publications, where the original text may be placed a) on the same page as its translation b) on the neighbouring page c) a number of pages later in the book. In the first case, the page number is given only once, rather than after the quote and its target language version. The same applies to one other source: Kossak (2001), which is not a bilingual publication but page numbers exactly correspond in the Polish and English editions.

3. The original texts are quoted with all the errors that they contain. Corrections are suggested only for those mistakes that distort the view of the natural world. The remaining ones, no matter how serious, are largely ignored.

INTRODUCTION

This work has been inspired by the lack of publications on the difficulties the translator faces when confronted with texts rich in nature terminology. The other source of inspiration was the simultaneously observed increase in popularity of bilingual and multilingual publications about nature, whose reliability left much to be desired. I then asked myself what made so many translators distort the image of the world in the target language so often (in the texts I analysed the target text was either English or Polish). The search for the causes led me to an attempt to find solutions and recommendations, both general and specific, which would sensitise translators to the problem and help them take decisions in particular translation processes. Hence the basic objectives of this book are to show that:

1. the main hindrance in the translation of nature terminology is differences in the categorisation of the world in various languages.
2. another obstacle is the multitude of ways in which the world can be categorised within one language, which is brought about by the existence of numerous scientific disciplines, schools of thought, changeability in time, overlap of specialist and folk categories[1] and by the resultant necessity of intralingual translation.
3. awareness of these facts helps translators avoid errors.
4. this awareness can be raised, e.g. by the text analysis based on cognitive linguistics, in which equivalence is established on the level of imagery. This means that potentially equivalent terms are compared in respect of the degree of their conventionalisation and dimensions of imagery: salience, schematicity and perspective[2].

---

[1] Despite voices calling for making a distinction (Habrajska 1996: 223) between classification, as referring to scientific activity, and categorisation as connected with the folk experience of the world, this work does not confirm the existence of clear-cut boundaries. Following Jolanta Maćkiewicz, a Polish scholar particularly interested in scientific and non-scientific categorisation (see Maćkiewicz 2004, 1996, 1992 and 1991), I will be using both terms interchangeably.

[2] I use Langacker's terminology, presented in a simplified manner in 1.3.1. below, pp. 15-16.

5. up-to-date extralinguistic knowledge is of great significance for the translator. It is indispensable to determine whether potentially equivalent terms are really equivalent and whether the categories they name have similar borderlines, membership and prototype.
6. it is imperative to approach dictionary equivalents with caution. It concerns both general and specialist dictionaries as they contain errors and offer a range of possibilities without any indication which are obsolete or rejected as inadequate in view of progressing research.
7. specialist texts include non-specialist terminology and vice versa, therefore it is important for the translator to examine terminological equivalence on the level of imagery irrespective of the type of text in which they appear. Errors in the use of nature terminology are visibly ignored, especially in belles-lettres, where, though sometimes irrelevant, they equally frequently lead to the distortion of the world presented in the book (e.g. when a birder claims that he can hear a moorhen on the canal (the moorhen cannot swim), or when the translator makes zoologists look for sousliks (which live in open landscapes and are unable to climb) up in a giant tree in the rainforest.

Following Putnam (reported in Muszyński 1996: 40), I have assumed that scientific terms concerning nature comprise a more precise 'description' of reality and as such they deserve propagating, as opposed to folk terms. While this assumption does not seem to be controversial in reference to informative texts, in belles-lettres it is valid in the cases justified by the context. In each case I try to prove it in my analyses.

I have also assumed that in order to show the complexities of categorisation it is necessary to collect a large database. The choice of terms stems from the analysis of the greatest number of Polish non-specialist publications translated into English I have managed to find, specialist texts mainly on phytosociology, especially those concerning the EU Natura 2000 programme, and translations of novels and other narratives written by several Polish and British authors (e.g. John Fowles, Gerald Durrell, Czesław Miłosz, Jarosław Iwaszkiewicz), which abound in nature lovers and experts, so also in relevant terminology. My analysis has yielded a rather diverse set of terms that seem to cause translators a lot of trouble. They all concern the natural world understood as both animate and inanimate nature affected by forces independent of human beings as well as by human activity. In my search I have been limited by my personal knowledge and verification abilities – hence favouring phytosociological and ornithological questions. All comments are based not only on my intuition but also on almost eighty specialist sources and consultations with specialists – biologists and environmentalists from the University of Warsaw and the Institute for Land Reclamation and Grassland Farming at Falenty. Although I have not carried out questionnaire research into what images readers of the original and the translation tend to have behind the terms examined, my interpretations of their possible reception stem from the fact that the dictionary definitions reflect the frequency of usage research, and that the data I have gathered give insight into typical images in the mind of numerous translators.

I concentrate on lexical problems, mentioning grammar only where it is difficult to separate both of them without engendering confusion. I do not intend to study 'fundamental concepts going to the heart of our conceptual systems: space, time, causation, event structure, aspect, evidentiality, fundamental classification of objects' (Lakoff 1987: 329), but 'superficial conceptual differences [such as] vocabulary, specialized terminology, words for obscure concepts' (ibid.: 328-29). Although we can agree that 'concepts that are spontaneous, automatic, and unconscious [those that are grammaticized or metaphorised] … simply have a greater (though less obvious) impact on how we understand everyday life than concepts that we merely ponder' (ibid.: 335), superficial, i.e. lexicalised conceptual differences, pose enough translation problems. Besides, among concepts that we 'merely ponder', such as those expressed through vocabulary, a gradation of spontaneity, automaticity and consciousness in use is observed. Short words can be juxtaposed against long cumbersome labels. The existence of a single word to denote a conceptual category bears evidence to the importance of a category for a language community (see Maćkiewicz 2004a: 53 or Dougherty 1993: 163-164).

In Chapter One: *Terminological and cognitive approaches to the categorisation of nature* I come to the conclusion that terminology is a product of categorisation, which is context-dependent. In translation context-dependence is particularly obvious, therefore it is necessary to analyse terminology within a cognitive framework rather than within the methodology used in terminological studies. The cognitive approach is adopted for a number of reasons: 1) finding the middle way between extreme universalism and relativism; 2) seeking to eliminate the oppositions dictionary/encyclopaedia and synchronic/diachronic studies (the appreciation of context, not just co-text); 3) ascribing an important role to categories in our thought organisation; 4) viewing categories as relative entities organised around prototypes rather than having a set of shared properties which are sufficient and necessary for a member to be included.

One of the reasons is given more prominence: categories are proved to show prototypical effects and are characterised by (E. Rosch's nomenclature): background framing, fuzziness, gradability and stretchability. Such a view of categories allows us to account for the observation that boundaries between specialist and non-specialist categorisation are blurred. Three cognitive theoretical frameworks for translation studies are analysed: Hejwowski's, Gutt's and Tabakowska's with a view to applying them selectively in my own analysis of nature terminology in translation. Combined with these is another theoretical framework – the linguistic view[3] of the world developed by Bartmiński, Tokarski and others. Unlike them, I do not try to provide multi-faceted descriptions of the view of the fragment of the world but rather concentrate on what interpretation or image a particular term projects in a given context.

In Chapter Two I present the categorisation of nature in Polish and English specialist texts. Contrary to all recommendations and idealistic visions[4] features of categories resulting from their prototype-oriented structure (see Ch.1.3) remain true for specialist writings, too. They are pinpointed as the main cause of translation problems. The role of background framing is enhanced since terminology is shaped by the author's (the interpreter's of the reality) scientific and practical interests and his/her cultural perspective. For instance, if a habitat is interpreted as *mechowisko* in the Polish documentation, it is possible to translate it into the Natura 2000 category montane and lowland alkaline fens in the form of flush vegetation sedge beds and moss beds, which guarantees its protection. If it is classified as *turzycowisko (szuwar wysokoturzycowy)*, it cannot be matched with the above Natura 2000 category and is stripped of protection. The chapter discusses intralingual and interlingual translations of woodland habitats (such as *świerczyna na torfie, dąbrowa świetlista, alder swamp forest, bog woodland*) and non-forest habitats (such as *murawy napiaskowe, łąki trzęślicowe,* raised bog, fen) from the language of foresters into the language

[3] This is one of the three possibilities for the English name of the framework. The others I have met are: 'image' and 'picture'. With regard to Grzegorczykowa's definition of *językowy obraz świata:* 'the conceptual structure characteristic of a specific language with the help of which people who speak this language classify/ interpret the world' (2004: 45) it seems that 'view' is the best choice. Since, however, the English summaries of the articles published in the volume *Językowy obraz świata* (Bartmiński ed.: 2004) consistently use 'image', I follow them when referring to this source. Sometimes I write 'picture', as it appears frequently in the relevant literature (cf. Bartmiński and Tokarski (eds): 1998 or Zinken: 2008).

[4] See e.g. Szujecki (1998: 29-30), Felber and Budin (1994: passim).

of nature conservationists as well as attempts to correlate names of species which vary regionally, historically and according to different taxonomic systems. I also present controversies within the same discipline over such terms as mire, which some authors consider a subcategory of peatland, while others as a category superior to it.

Chapter Three details the categorisation of nature in Polish and English texts for non-specialists. It is observed and exemplified that non-specialist texts include a lot of specialist terminology and it is difficult to determine the boundary between what is specialist and what is not. The view of the world emerging from them never has just one perspective. Examples (*barć*, *bór/las/puszcza*, *grąd*, *łęg*, *ostoja*, *szuwary*, *wrzosowisko*, buntings, copse/coppice/grove, game birds, midges, waterbirds/waterfowl/wild fowl, etc.) are analysed in terms of a degree of departure or similarity of category members from the prototype, encapsulated in an equivalent expression in English and Polish at the level of: 1) conventionalisation (unequivocal imagery, gestalt perception vs. visualisation with effort); 2) salience; 3) schematicity (schematic presentation vs. instantiation); 4) perspective. According to Tabakowska, who follows Langacker, these are components of the level of imagery at which all translations should be analysed in order to aim at the best quality[5]. The shifts of imagery are also considered for onomatopoeia (mainly the notation of animal calls) and in the case of puns using nature terms.

In Chapter Four the concept of 'linguistic view/picture of the world' is called upon again in order to show how translation errors give a distorted linguistic view of the world. Firstly, an attempt is made to use Hejwowski's typology of errors for various examples of mistranslated nature terminology (among the types are: unjustified borrowings and domestication, calques, overspecification, underspecification (using hypernyms), etc.). This is followed by an error analysis based on abuse of such features of categories as stretchability or fuzziness with regard to a particular background framing. In each case the result is the wrong image. The whole series of wrong images contributes to the distorted view of the world.

The whole work aims to enhance the significance of the didactic mission of translation, in this case reflected in promoting a more scientific view of the natural world through terminological accuracy. Translators must remember that there is nothing degrading in assuming that the readers of the target language text expect the natural features of the source language world

[5] In contrast to the application of Langacker's concepts in Tabakowska (1993), in this book it is sketchy and subservient to its other objectives.

to be different and therefore worth learning about, insofar as gradual learning about the diverse world around them is worth their effort.

Thus, translation must be used to 'teach' new concepts or new categorisations of concepts to representatives of a foreign culture by applying correct lexical labels. Eva Hoffman's experience seems to validate such a claim. In *Lost in Translation*, after describing the pains of learning to rename objects she knew from Poland in the language supposed to become her new 'mother tongue', she breathed a sigh of relief when taught names of flowers that she had not formed any concepts of in Cracow. She notices '[f]or now, there are no Platonic azaleas, no Polish hyacinths against which these are compared. The language of this is sufficient. I am here now' (Hoffman 1998: 280). If so, and if we take into account the fact that modern people are increasingly less familiar with the surrounding animal and plant life, it becomes self-evident that when a publisher decides to release a book or an album in which nature is given a degree of prominence, the chance to be truly educational cannot be disregarded by the translator, who happens to attach more value to, for example, high technology or literary theory. Unfortunately, as the corpus collected for this study shows, the chance to feed the knowledge-hungry readers is often ignored. Postulating the translator's didactic responsibilities, I have a sense of educational mission myself, which in this book acts as a stimulus for including rich exemplification in the chapters dealing with the images of the natural world. Yet, in my proposals of corrections for mistranslated passages from belles-lettres I am never guided by orthodoxy, primarily looking after the text's logic and coherence rather than terminological precision at any cost. Suggestions of practical solutions are meant to help translators as I subscribe to Newmark's opinion: 'Note that translation theory is concerned with choices and decisions, not with the mechanics of either the source language (SL) or the target language (TL) … . As with much literature *à thèse*, the examples are often more interesting than the thesis itself' (Newmark 1988: 19).

However, the ensuing pages are not simply thronged with examples for the purpose of impressing or distressing the reader. What at first glance, especially in Chapter Two, may seem an exaggeratedly ample exemplification of but one problem is, actually, an elucidation of its innumerable aspects. It must be immediately noticed that these are not purely linguistic aspects, as they can barely be separated from extralinguistic ones. If the data provided give an impression of overabundance, one must realise that they touch upon a mere fraction of possibilities. Overwhelming diversity and perpetual changeability underlie the categorisation of the world. As categories need to be translated from discipline to discipline and from language to language, the same, unavoidably, plagues the translation process. A study of translation cannot responsibly claim that such complex matters do not require detailed, often interdisciplinary, exposition.

CHAPTER ONE

TERMINOLOGICAL AND COGNITIVE APPROACHES TO THE CATEGORISATION OF NATURE

1.1. A non-terminological approach to terminology

While the bulk of the work concerns terminology, it is not strictly terminological. Translation problems cannot be explained only through the tools offered by terminology, understood as a discipline investigating relationships between terms and concepts. This is because:

> [t]ranslators and terminologists work in quite different modes, … [t]ranslators generally attempt to match the largest possible units of meaning as they arise in a specific text; terminologists have no interest in temporary and casual collocations of concepts as they are brought into a particular relationship by a writer. Translators work with concepts and terms in context; terminologists isolate terms from context and fit them into an abstract system of concepts. (Sager 1997: 251)

It must be immediately stated, though, that terminological studies come in handy as they explain why so much intralingual translation must be done before interlingual transfer begins. Besides, whereas theoretically translators 'move from the linguistic item to the concept in order to find a corresponding linguistic realization in another language', in practice, like terminologists, they often 'need to establish identity among the concepts' (ibid.: 259) first, and having done so, find or create the matching linguistic item. The terminologist's role consists in trying to achieve an ideal by an adequate formulation of terms and imposing standards for everyone to use. The translator should help impose standardised terms to the degree contexts allow them to do so. But s/he does not work within the ideal world. All the time confronted with a messy reality and a wide variety of recipients, they have to make compromises over various expectations. As Sager, observes '[t]he greater the differences between speakers the greater the need to use a higher proportion of definitions, paraphrases and circumlocutions rather than standardized terms' (ibid.: 258). What is more, there must be an agreement among recipients on the desirability of the relaxation of 'the aesthetic, emotive and other non-informative requirements of language in order to enjoy the greater efficiency created by standardized reference' (ibidem). Terminologists draw our attention to certain facts of significance to the translation process:

1) there is general language and special languages equal to the number of scientific disciplines;
2) main differences between these two occur in their lexicon;
3) scientific disciplines assign diverse concepts to the same objects and classify concepts according to varied criteria, which give rise to rich terminology;
4) concepts hidden behind superficially identical or similar terms can be different across languages;
5) there are terms which function in specialist contexts as well as general ones, but have acquired different meanings in each;
6) such terms in general language tend to be polysemous;
7) there are international norms and requirements concerning terms and nomenclature specifying, for instance, when it is better to assign a new concept to an already existing name, or that new names should be short, otherwise they will be abbreviated anyway, engendering ambiguity (cf. Felber and Budin 1994: 150-51).

From the translator's point of view it is not important whether we ought to talk only of specialist terminologies or of specialist languages (sublanguages of the general language). Neither should s/he worry at what point general language becomes specialist language (cf. Pieńkos 2003: 258-263). For practical purposes it suffices to remember that there are texts, which J. Delisle calls pragmatic, and which serve mainly to inform, with an aesthetic aspect assigned a secondary role (cf. Pieńkos 2003: 271). In such texts terminological precision cannot be sacrificed so as to satisfy the recipients' taste for stylistic or aesthetic sophistication. Nevertheless, the translator should remain alert to features of a particular specialist text type such as conciseness, preference for nominalisation and passive voice, etc. The translator cannot forget that s/he stands no chance of escaping terminological issues even if s/he is preoccupied with belles-lettres. In fact, the division between literary and specialist texts is artificial since terminological problems concern literary texts too, and there they are, actually, most difficult to solve (cf. Pieńkos 2003: 242-43, 91 or Hejwowski 2004a: 249). Although some translation theorists, like Gutt, seem to justify the use of imprecise and erroneous terminology if for the projected recipient it does not seem to be optimally relevant (Kasprzak 2008: 181), others support terminologists in their struggle for terminological order. For example, Pieńkos (2003) advocates a thorough and comprehensive education of translators because they are responsible for keeping the best standards to the degree that they should correct texts in which 'linguistic and terminological anarchy coupled with unclear and incoherent argumentation rule' (2003: 282-84). Hejwowski (2004b: 142) condemns errors made by translators who have failed to inquire about certain general or specialist facts, notably names of animal species. In his book, *Translation: A cognitive-communicative approach*, he includes two quotations, which best summarise the attitude endorsed in the present work. The first comes from Lehrer:

> If the typical readers do not have aspens or elms where they live, should a translator select a different tree? I would say 'no' for these texts and for other works of fiction, since a literary translation should be aimed at an ideal reader – one who is knowledgable about the source culture, rather than a typical, less sophisticated 'average' reader. (Hejwowski 2004a: 107)

In the second, by R. T. France, the word 'author' can be easily replaced with 'translator': 'It is a poor author who aims to communicate only with the lowest common denominator of his potential readership' (ibid.: 109).

Attention that translators ought to pay to both scientifically defined and folk terms inevitably guides them to the term's referents in the shape of concepts. Remarkably, most problems originate at this level since, as Wilhelm von Humboldt underlined, 'each language describes and expresses, in a way peculiar to itself, the world outside, creating their own image of this world. Therefore each language presents its own interpretation of the world' (Pieńkos 2003: 232, translation mine). In effect, translators often fall into a trap trying to render a term in the target language, while the real issue is to find 'a linguistic equivalent of a concept or phenomenon' (ibid.: 285). Pilegaard (1997: 165) puts the same idea in the following way: '… in practice, translators should always check the comparability of the concepts at issue "by looking closely at the words people use"' and stresses after Alexander and Slater (1987) that it is 'the common fallacy that scientific discourse in general and medical discourse in particular is subject to straightforward objectivity'. An important related question is that languages determine a term's denotation in different ways. For instance, French *outillage* covers the meaning of the English 'tool' and extends beyond it, embracing also measuring and holding devices (Felber and Budin 1994: 86-7) or the German *Schlagball* and the English 'cricket' share the equipment used in the game but have dissimilar rules (ibid.: 118). Pieńkos adds that each word (term) has its own 'usage zone', 'sometimes different in a literary text than in a scientific one, where the boundaries between zones are fuzzy and move here and there from time to time' (2003: 277). He emphasises that terminological differences, though to a lesser extent than in the social sciences and humanities, 'exist within exact sciences and engineering, but they are less obvious and more subtle' (ibid.: 270). He also warns that translators may encounter 'terminology, which is nothing but the author's *ad hoc* creation bearing features of professional jargon' (ibid.: 244). This is something on which conscientious translators waste a lot of their time, looking for established equivalents of what has not been established yet at all.

The most interesting aspect of the terminologist's work, with implications for translation, is the examination of 'denotational relationships within scientific vocabulary, which occur simultaneously in many disciplines and often outside them, i.e. in general language [which according to Knobloch and Schaeder (1994) are between]: 1) a word and a term, 2) the meaning of the word and the concept 3) the language of teaching and the scientific language and 4) the encyclopaedia and the dictionary.' (Pajewska 2003: 31). Pajewska goes on to enumerate such examples of Polish words as *analogia, kompleks, model, proces, struktura, system*, etc. As they resemble English words, no translation is necessary. However, it must be stipulated that their senses and usage do not overlap in both languages. Apart from such extremely multi-purpose words there is a much wider range of those with a narrower usage but exploited throughout general language and one or two sublanguages: *bagno, uroczysko* or *łęg*, which will be discussed *in extenso* in Chapter Three.

The complex relationships sketched above result in such phenomena as synonymy, homonymy, polysemy, false friends, etc., which plague translators. Terminological studies assist in overcoming them through the publication of good specialist dictionaries. However, such dictionaries usually present the ideal, context-free options. Theoretically, in contrast to general language dictionaries, they understand synonyms as terms that can be substituted for each other in the same context. Practically, inconsistencies can creep in and synonyms may become 'words with a similar meaning'. In general, synonymy is regarded as undesirable in them. Whatever the wishes, such postulates of terminologists as unequivocality of meaning and contextual autonomy of terms often remain postulates because of the multiple naming of the same concepts and focusing on different features of the same referent in different specialist texts and because of conflicts between everyday uses and specialist uses (see Pajewska 2003: 110-111).

Pieńkos goes as far as to talk about pathology in terminology. Consisting in ambiguity, it stems from 'the [uncoordinated] creation of concepts and their systems in different civilisational and cultural circles' (Pieńkos 2003: 227) and 'the lack of unanimity among professionals' (ibid.: 316). Robinson illustrates this situation with his search for a Finnish equivalent of 'flip-off seal' needed for a translation of instructions for a pharmaceutical product. He obtained three very different but confidently formulated answers from three medical doctors, of which one sounded too colloquial and one overextended in meaning. Upon such findings he reflects: 'This is a good reminder that even specialists belong to more than one community; and even within one community they often maintain two or more registers, one technical and "official", one or more slangy and informal' (1996: 201).

In conclusion, for translation purposes terminology must be studied in context: 'The understanding of terms relies not only on familiarity with the definitions of concepts that they stand for but also of the texts they are used in. ... [T]erms function differently in strictly scientific texts, or in popular science or didactic publications' (Pajewska 2003: 52, translation mine). However, the context cannot be limited merely to the co-text. It should be the broadest possible, encompassing the user's knowledge, experience, views and any mental processes likely to take place in the minds of both the author and the recipient of the text that comprises terms. This necessitates looking for a cognitive approach to translation – in our case, to the translation of nature terminology.

1.2. The cognitive approach

A cognitive model has been chosen to account for translation problems primarily because, as Hejwowski puts it: 'Producing a text, translating a text and reading a text are all basically mental processes. Therefore in our opinion they should be dealt with by a cognitive theory, and not by one which limits its scope of interest to texts only' (Hejwowski 2004a: 258).

Another concisely formulated set of reasons can be repeated after Tabakowska:

> CL [Cognitive Linguistics], with its objective to establish probabilistic tendencies rather than deterministic rules, and with its pragmatic rejection of the economy of description and reductionism as basic methodological principles, opens new perspectives. It offers a theoretical framework for a more adequate definition of translation equivalence, and thus makes it possible to formulate some normative principles for translation assessment. It provides additional evidence for the claim that a "theory of translation" does not, and cannot, exist as an autonomous entity, in any absolute sense.' (1993: 32)

More specifically, cognitivist research has thrown some new light on one of the mental processes of great significance to translation – categorisation. Principles of categorisation are placed among general principles of cognition. It also explains why translation is possible even when conceptualising systems of two languages are very different by enhancing a role of universal basic experience and human conceptualising capacity. In other words, it enables us to draw optimistic conclusions concerning translatability, impossible with the adoption of the extreme version of Whorfian relativism. This version leads to a conclusion, protested by many authors[1], that translation is impossible – which runs counter to facts. At the same time, over-optimistic universalism is cut down to size. As Wierzbicka points out, Whorf exaggerated, because for effective communication we can always resort to paraphrases and circumlocutions in translation. But she hastens to underline that it can only be done at a cost – by 'using longer, more complex, more cumbersome expressions than those which we can use relying on the habitual ways of speaking offered to us by our native language' (1997: 7). Besides, cognitivism has tried to lift the opposition dictionary/encyclopaedia, placing emphasis on context-dependence and the role of extra-linguistic knowledge in linguistic studies, which underpins the line of argumentation on the deceptiveness of reliance on dictionary knowledge developed in the present work. The opposition between synchronic and diachronic studies has vanished too, which enables explanation of certain difficulties in translation, like those with *gajówka* (see pages 60 and 137 below) or *jodła/świerk*[2] (Ziółkowska 1983: 139). In both oppositions staking out demarcation lines is virtually impossible. The futility of looking for the limits of knowledge that the

---

[1] See e.g. Hejwowski (2004a: 14-15), Korzeniowska (1998: 71), Newmark (1988: 72-73).

[2] In 19th century Poland, north and east of Warsaw, the fir (*jodła*) bore the name which is today given to the spruce (*świerk*). In turn, the spruce was called 'fir'. Frequent mistranslations of these names can be traced back to this historical-socio-linguistic fact.

translator should possess is most vividly illustrated, e.g. in Tabakowska (1999: 54-60). In turn, the inevitability of intertwining synchronic and diachronic, linguistic and extralinguistic data is confirmed in a number of works on specialist languages and linguistic images of the world[3]. Pajewska concludes that derivational processes and tendencies observed in them as well as the representation of given thematic vocabulary are connected with the changes which occur outside language. Although in her work she focuses on diachrony, she notices that 'it is necessary to take into account aspects of synchronic description, because language changes, i.e. diachronic phenomena are eventually a result of the functioning of language – synchrony, which is always a start-off point' (Pajewska 2003: 215, translation mine).

Cognitivism as depicted by Lakoff is optimistic when it comes to translation in that it separates understanding from translation: 'Translation can occur without understanding, and understanding can occur without the possibility of translation' (Lakoff 1987: 312). This is connected with the relativisation of the commensurability of conceptual systems. Unlike Whorf, who claimed that if conceptual systems of two languages are radically different, they cannot be compared in any way, Lakoff is able to distinguish five criteria by which the systems can be compared. If they are incommensurable by one criterion, they may be commensurable by another. All depends on the formulation of the question. Commensurability is a gradable category (ibid.: 322). The main implication for translation is that translation is possible to a greater or lesser degree, or if it is impossible to translate one language (let us keep here to Lakoff's formulation, acknowledging Hejwowski's (2004a: 112) insistence that one does not translate languages but texts) into another sentence by sentence, preserving truth conditions, we can rely on the criterion of understanding or use, or framing, or organisation. To put it in a more straightforward way, language is not everything, it does not reflect our thoughts fully and equivalently. It is 'a convention which may, but does not have to, be an instrument of our thoughts' (Lewandowska-Tomaszczyk 1996: 67, translation mine). It is only a choice of certain phenomena, a huge mass of which escapes human cognition (cf. Anusiewicz 2004: 282). In other words, our language offers only an approximation of what our conceptual systems store and so do translations from one language into another. Once it is realised, demands concerning equivalence become more realistic. Lewandowska-Tomaszczyk calls cognitivism, as represented by Lakoff, Langacker, Fillmore and others, 'enlightened' relativism, which is universalist

[3] E.g. Maćkiewicz (1991), Marczewska (2002) or Pajewska (2003).

in their attitude to the cognitive base of all people, but assumes the existence of concepts specific to a particular language. Lakoff notices that in the case of concepts such as 'the basic level perception of physical objects' or ' "kinesthetic image schemas": structured experiences of vertical and horizontal dimensions, balance, inside and outside, and many others' translation 'is often possible, if not immediate' (1987: 312). But then he turns to concepts which pose difficulties in translation and describes their range as 'enormously wide'. This points out to a more pessimistic side of cognitivism – a tendency to find cases of untranslatability. However, as elaborated in section 1.3. below, the same cognitivist premises can give rise to more optimistic views on equivalence and translatability.

Analysing nature terminology within the CL framework requires acknowledging that human categorisation is organised around prototypes. As Lakoff points out, 'most of our concepts concern categories not individuals' (1987: 370). A single word does not label a single concept but a category of concepts within which there are more or less typical members. Our lexicon is organised in the same way as our general cognition. There are basic-level categories, which are not indivisible, like semantic primitives, and have a complex structure. Thus 'dog' is more basic than 'dachshund'. Paradoxically then, what is structurally more complex is cognitively simpler, perceivable as gestalts – overall shapes. Categories do not objectively exist in the world but are most often 'a result of human imaginative capacity' (Lakoff 1987: 309). Lakoff (1987) devotes a lot of space to explaining why the objectivist view of categorisation is erroneous. Most simply, the idea that there is a single right taxonomy is contradicted by the very abundance of taxonomies. The classical, objectivist (Aristotelian) view stems from the folk conception of the world, according to which there exist natural kinds – categories of things objectively existing in the world. Lakoff admits that many folk natural kinds superficially have well-defined boundaries and seem to be unquestionably real – fish, goose, horse, etc. But the problem emerges when the objectivist criterion for being in the same category is considered – members of a given category must share common properties. There are certain necessary and exhaustive conditions they must meet to belong to one category, otherwise they are excluded. Evidence from biological taxonomy demonstrates, however, that the decision which properties are to count is not made according to a standard, 'independent of human interests and concerns, that can choose between them and provide a unique answer' (ibid.: 189). Due to three different views on what data should be taken into consideration to determine the division into species, represented by cladists, phenetisists and evolutionists, this division can be different. If we accepted, following the objectivists, that there is only one correct categorisation scheme, and that this is the one advocated by cladists, such categories as zebras or fish would have to be deemed non-existent. There are species which phenetisists group under the common 'kind' zebra, or fish but cladists, using 'historically derived characteristics' rather than 'overall similarity' criteria, put into separate categories. Mayr's 'biological species' concept[4] tries to combine the two standpoints but it is not unanimously accepted, either. His findings confirm what in the field of cognitive psychology was discovered by Rosch, Kay, Kempton and others – that categories show prototypical effects. They have no homogeneous internal structure and clear-cut boundaries and they are defined relative to other groups. Certain folk categories regarded as objectively existing in the world correspond to scientific classifications, but some others do not correlate at all or they do only to some extent.

That categorisation is not uniform is easily instanced. What remains debatable is whether the world is objective in its organisation but humanity has no access to its objectively existing categories or whether some categories objectively exist in the world, while others are formed subjectively by people in different ways depending on their perspective. On the assumption that we have no access to objective categories, the diversity of taxonomies is explained as the outcome of our various attempts – not infrequently unsuccessful – to discover and name them. The world is well-structured irrespective of our efforts to make sense of this structure, which are reflected in all sorts of classifications and the resultant labels. Lakoff takes the second stand of those juxtaposed above. He opts for, what he calls, experientialism, underpinned by basic realism, which acknowledges, *inter alia*, 'the reality of a world existing independent of human beings' and 'constraints on our conceptual systems due to the nature of that reality' (1987: 266). This allows him to reject total relativism, at the same time preserving essentially relativist views. He seems to bridge relativism with objectivism successfully[5]. Although an admittedly attractive proposal, it does not appeal to everyone. Muszyński dismisses it as 'trying to join opposite river banks with a bridge constructed along one of them' (1996: 38, translation mine). In the present work, taking one of the two positions described above is not necessary. Interest here lies in the fact that people do categorise in different ways across cultures or social divisions, and that even 'a given person may hold

[4] For more information see Chapter 2.4.
[5] See e.g. Grzegorczykowa 1996: 11-26.

one or more folk theories and one or more expert theories in areas like medicine or economics or physics. It is commonplace for such idealized cognitive models to be inconsistent with one another' (Lakoff 1987: 121). Such diversity poses enough down-to-earth problems in need to be solved by the translator.

Putting philosophical deliberations aside, attention will be paid here to the palpable reality of prototypical effects in categorisation. These have been confirmed in many studies, four of which are discussed in more detail below. Dougherty (1993: 160-185) emphasises that the basic level in a particular classification depends on the significance of a given life sphere to a specific group of people. In the spheres of little cultural significance the basic level is moved towards more general distinctions. More general categories become most relevant, while those more specific cease to be widely recognisable 'gestalts'. American city dwellers draw distinctions at the level of life forms ('bush' vs. 'tree') rather than at the level of genera ('pine' vs. 'fir'), let alone species ('ponderosa pine' vs. 'lodgepole pine'). Native American tribes, for whom folk botany is a matter of everyday interaction and, in fact, of survival, establish the basic level of classification lower in the hierarchy. For them it comes naturally to distinguish a vast number of particular species. Dougherty goes as far as to imagine human communities settled somewhere in Outer Space for which the basic level of botanical classifications could shift upwards even in relation to American townspeople. Potentially, the contrast bush/tree would become abstract enough for them not to be immediately perceivable. Instead, they would limit themselves to plant/animal differentiation only. All in all, the basic level of categorisation is a relative issue rather than stemming from the objective structure of the world around.

Likewise, Hunn (1993: 73-96), also analysing the folk botany of Native American tribes, concludes that only some life forms are natural entities, distinguishable by morphological resemblance. Most of them, however, are grouped under common names for reasons other than overall similarity. These are usually very practical and understood by outsiders only when questions such as: who, when, why and how? are answered. It happens that certain plants, morphologically similar and scientifically classified in the same taxonomic unit, belong to different categories in folk botany because, for instance, one is gathered by men in spring with the help of a hoe in order to be eaten raw, and the other is collected by both sexes in summer with bare hands and then baked in an earthen oven for three days. Moreover, folk taxonomies are full of gaps – a lot of organisms bear no names, being referred to as 'something similar to ...'. They form an artificial periphery, existing for a given human community, not in nature. They are residual species without practical importance and thus without specific names. Scientific taxonomies are not complete either but they try to classify organisms within a hierarchical structure based on morphology and genetics rather than on the core/periphery opposition derived from a mixture of morphological observations and associated cultural practices (or lack of them).

American observations have found confirmation in Polish studies reported in: Szczekocka-Augustyn, Wereszczyńska, and Zagrodzka (1996) and Habrajska (1996). Szczekocka-Augustyn, Wereszczyńska, and Zagrodzka (1996) examine fuzzy categories. Having tested 500 students and 300 lexemes of the Polish language, they conclude that the boundaries of higher-rank categories arbitrarily created by human beings, such as those of artefacts (tools vs. devices vs. instruments vs. machines) are least clear-cut in comparison with the higher-rank categories of animals (mammals vs. birds vs. reptiles, etc.) or plants (trees vs. bushes vs. flowers). In some cases such categories even lack typical exemplars constituting their centre. Although statistically animals and plants tend to be grouped under the same categories more consistently, half of the people tested confuse reptiles and amphibians, and a lot of them liberally mix up fruits and vegetables. Habrajska describes a test in which testees drew a list of 310 lexemes denoting plants and were then asked to arrange them in categories. In effect, a large number of categories were created, for example, cultivated plants, pot plants, cereals, herbs, garden plants, field plants, edible plants, weeds, root plants, vegetables, flowers, fruit plants, etc. They were characterised by free profiling, incompleteness and inconsistent criteria of selection, so they often overlapped. These folk categories were compared with scientific ones, which have a hierarchical structure and whose sets are supposed to be exhaustive and mutually exclusive. Most importantly, the author postulates the introduction of separate terms for folk and scientific ways of compartmentalising the world, the first – categorisation, the second – classification. As shown later in the present work, mainly in Chapters Two and Three, this is an ideal situation, contradicted by reality, in which scientific categories become as subjective and fuzzy as the folk ones due to a multitude of scientific classifications in circulation and efforts to adopt them for non-professional uses.

Generally, both early cognitivist psychological experiments and later linguistic studies indicate that categories show prototypical effects and, following E. Rosch's nomenclature (reported in Kardela 1996: 23-24), are characterised by:

- background framing
  (category formation depends on the encyclopaedic knowledge, experience and views of a language user)
- fuzziness
  (there are no clear-cut boundaries of the meaning of words or terms, categories denoted by them are perceived as gestalts)
- gradability (even when the boundaries are well-demarcated there are better and worse category members bound by what Wittgenstein calls 'family resemblances', with higher or lower 'cue validity', to use Rosch's term)
- stretchability (a category can include elements of a completely different category)

In this work, these properties – valid for both folk and scientific categorisation – serve as a start-off point for a discussion on what the translator can do in order to avoid mistranslation of nature terminology. In other words, how s/he can detect cases of 'overfuzziness' or 'overstretching' in the original, and how to prevent such cases from permeating into the translation.

1.3. Translation within a theoretical framework

1.3.1. Cognitive accounts of translation: Gutt, Hejwowski and Tabakowska

According to Hejwowski (2004a), within the cognitive framework three theories of translation have been developed: Gutt ([1991] 2000), Tabakowska (1993) and Hejwowski (2004a). They differ significantly. The newest of them is by Krzysztof Hejwowski, who, acknowledging a number of merits the other two have, identifies in them various drawbacks, which have inspired the elaboration of his own theory. Although Gutt's relevance-theoretic account of translation gives due weight to mental processes involved in translation, it aspires to be reductive. Neither Tabakowska nor Hejwowski believe one can explain such complicated processes as communication – and, consequently, translation – with the help of a one all-encompassing principle. Gutt defines translation as 'interlingual interpretive resemblance' (2000: 196), understood not as 'an exact duplication of thoughts in communication' but as 'enlarging mutual cognitive environments' (ibid.: 213). This can be achieved with 'the guidance provided by the search for optimal relevance, taking into account the specific contextual background of the audience s/he is working for' (ibid.: 237). The search for optimal relevance means that communicative effects yielded should match the processing effort the given audience is prepared to invest in the particular context. This has an important consequence: '[t]he translator's or interpreter's task is to ensure that the receptor language text he comes up with ".... is the most relevant [ostensive stimulus] the communicator could have used to communicate" (1986a: 158)' (ibid.: 230), or put in other words: 'translation can only be expected to be successful when processed in the intended context' (not just co-text) (ibidem). If so, the decision as to the intended context belongs to the translator since translation is not an act of direct face-to-face communication contexts and, in effect, receptors with their expectations are bound to change and cover a broad spectrum. What flows from this is that almost any decision taken by the translator can be easily justified. Taking an easy way out, s/he may decide that the most likely target readers of a novel are those who seek entertainment or want to kill time on a long journey. On such an assumption, terminology from various professional fields that the author has smuggled into descriptions or dialogues can be either omitted or rendered inadequately, even if it entails lack of coherence and illogicalities in the plot. Allusions to high-brow literature or art can share the same fate. The demanding, sophisticated reader is ignored and his/her processing effort actually increased by repeatedly exposing him/her to illogical, inconsistent or otherwise deficient fragments of text. Permissiveness of that sort bodes ill for the future of translation and sanctions a lot of substandard work done so far. Besides, a lot of decisions translators have to make are compromises, which – by always satisfying the sense of relevance of one type of receptors – increase the processing effort of some others[6].

The observations described above are shared by Hejwowski: 'This stress on minimal effort on the part of the recipient is very dangerous because it can encourage some translators to simplify translated texts, which is justifiable only in very special and very rare circumstances' (2004a: 26). He also writes that translation has much to do with some of the situations excluded by the Relevance Theory's definition of communication, such as overhearing and eavesdropping. Relevance Theory concentrates on acts of ostensive communication, whereas 'from the sender's point of view, TL recipients are not the intended addressees of many translated texts' (ibid.: 24).

Moreover, Hejwowski objects to Relevance Theory's minimizing the role of the recipient in communication, pointing to the 'effort after

---

[6] See e.g. Kasprzak 2008: 171-187.

meaning' principle formulated by Bartlett in 1932. This principle 'urges us to look for hidden truths, to get to the bottom of problems, to analyse and to ponder' (ibid.: 85). Although generally people are guided by this principle, the intensity of its application varies. However, by no means can we assume that in all acts of communication the effort initially invested is always sufficient to achieve adequate effects. We can be persuaded that before we give up reading a text that at first appears to be quite incomprehensible, it is worth making more effort; or before we conclude that there is nothing in it that enlarges our cognitive environment, we had better search for something that does, perhaps in a way subtler than expected. In other words, effort is not always adequate to contextual gains. We may be prepared to fail to gain enough but we continue reading or otherwise participating in a communicative act – 'utterances often require serious reconstructive and reparative effort on the part of the recipient' (ibid.: 29). Thus, Gutt's claim that the intended interpretation of an utterance is the one consistent with the principle of relevance and 'there is never more than one interpretation that fulfills this condition (1991: 31)' (quoted after Hejwowski 2004a: 28) – is untenable. As can be seen, Hejwowski takes issue with some core tenets of Gutt's account of translation. Yet, he calls it 'a major breakthrough in the field of translation theory because it clearly demonstrated that the translation process can only be described by means of modelling what happens in the translator's mind' (2004a: 30). It also fits Hejwowski's optimistic view on translatability stemming from the approximative character of human communication (cf. Gutt 2000: 213 quoted above, or Gutt 2004: 168). Communication dwells heavily on such basic human characteristics as 'the tolerance for imperfection and reluctance to be precise, willingness to compromise and to choose second-best solutions or the lesser evil' (Hejwowski 2004a: 75) – and so does translatability.

By contrast, Tabakowska (1993) uses the term 'untranslatability' quite frequently. Although two years later she ended her article "Językoznawstwo kognitywne w teorii i praktyce przekładu: oleodruk i symfonia na dwa fortepiany" ["Cognitive linguistics in the theory and practice of translation: a chromolithograph and symphony for two pianos"] on an optimistic note, likening the process of translation to an individual (not worse or better) interpretation of a musical score (Tabakowska 1995: 41), this translatability scepticism was noticed by Hejwowski and stimulated him to propose his optimistic cognitive-communicative theory of translation. Both Tabakowska and Hejwowski believe that translation equivalence is a continuum, i.e. there are better and worse cases. They also share the conviction that texts are translatable insofar as they are interpretable. Just as communicators interpret their respective utterances in various ways and no single objective meaning exists, we interpret translations – relative to a concrete situation and recipients. This thought is vividly expressed by Tabakowska:

> … disparity concerning inventories of domains, or cognitive structures, is not limited to interlinguistic relations, of which translation is one. In other words, in order to prove "untranslatable" a text need not be subjected to the operation of being turned from one language to another at all. (1993: 113)

Both Tabakowska and Hejwowski develop a text analysis method which allows them to compare the original and its translation on a number of planes, with a view to inclining the translator to reflect upon a broad spectrum of aspects which must be considered in the process of language transfer. Tabakowska regards the level of imagery as being most important and revealing. Following Langacker, she understands imagery as the language user's ability 'to construe a conceived situation (Tabakowska consistently uses 'scene' instead) in alternate ways – by means of alternate images – for purposes of thought or expression' (Langacker 1987: 110). Thus, imagery is what literary studies call style and what Jakobson labelled as 'poetics of grammar' (cf. Tabakowska 1993: 51). Taking advantage of Langacker's terminology, she distinguishes several dimensions of imagery:

1. comparison, due to which categories are formed through summary and sequential scanning of the field of vision.
2. selection (of domain, scale and scope), thanks to which useful information is retained, while the pieces judged irrelevant are rejected.
3. abstraction, used to establish the basic level of a category as opposed to subordinate and superordinate levels.
4. activation – depending on the conceptualiser's interests at a given time only some aspects of experience are paid attention to.

Owing to these dimensions imagery can be considered as a matter of degree of

- conventionalisation,
- salience,
- schematicity (specificity),
- perspective (Tabakowska 1993: 51-80).

The relation between the dimension of comparison and the degree of conventionalisation is as follows: categories we form can be more conventional or more idiosyncratic, hence they can have a varying degree of linguistic conventionalisation. Selection corresponds to salience[7], which, among other things, involves deciding which constituents of the entire conceptual content of an expression should be profiled and which belong just to the base – necessary to understand the concept; the decision concerns the choice of immediate, maximal or perhaps intermediate scope of an expression to be evoked. Abstraction relates to schematicity – one expression can be a very general schema, perceived vaguely as an overall shape and another can be its instantiation. Activation goes together with the degree of perspective, or vantage point from which a concept behind an expression is viewed, for instance, from inside, characteristic of common people, or from outside (bird's eye view), expected of scientists.

Hejwowski underrates the importance of imagery in text analysis for translation purposes by remarking that the reliance on it leads to exaggerated meticulousness and pessimism as to translatability. He criticises Tabakowska for being too text-oriented and over-concentrated on translation techniques because it is not texts but their mental representations that are supposed to be equivalent. He sees her as excessively focused on untranslatability despite the fact that she does not imply total untranslatability. She separates equivalence of content and equivalence on the level of imagery, in several cases declaring that translatability via compensation techniques on the level of content is possible, the problem being unavoidably mismatched imageries. Whereas he appreciates her sensitivity to text, which aims to raise standards and maintain them high, he regards the best available compensation techniques as sufficient, since, anyway, 'abilities to construe a conceived situation' depend on multiple factors shaping readers' reception, which after all are

---

[7] According to Langacker there are many kinds of prominence (salience). In this work I am mainly concerned with profiling. The profile is a substructure an expression singles out within its conceptual base as a kind of focus of attention. It is the expression's conceptual referent (designatum) (cf. 2009: 7). My analyses go sometimes further, looking at multiple domains invoked by a predication, which characterise different aspects of the profiled entity. That is because: '[s]emantic contrasts often hinge on the inventory of domains, as well as their ranking for relative prominence' (Langacker 1991: 62). Since I write mostly about nouns – expressions that profile things, I do not consider the second type of salience distinguished by Langacker: the trajector (primary figure) – landmark (secondary figure) organisation, which is valid for expressions profiling relationships – verbs, adjectives, adverbs and prepositions.

outside the text and out of the translator's control. Instead of one spacious level of imagery he suggests looking at a range of levels: lexical items and syntagms, verb frames, scenes and scripts, and schemata (including intertextuality, modality and genre) (Hejwowski 2004a: 115-116). In itself this dissection exercise does not change the fundamentals. All those ingredients essentially tally with what constitutes imagery, as they 'represent objects, actions, feelings, ideas, states of mind and any sensory and extra-sensory experience' (Cuddon, cited in Tabakowska 1993: 45). Mistranslation can be often said to produce undesirable pictures or notions in the mind of the recipient, irrespective of whether we choose to call them images, scenes, scripts and schemata, or solely images.

What does make a difference is Hejwowski's treatment of equivalence as a relative value. Text B is equivalent to the original Text A when: '... taking into account the translation difficulties and other translation solutions that we are able to envisage at the moment, Text B allows its readers to arrive at an interpretation that is relatively close to the one that was presumably arrived at by the readers of the original' (Hejwowski 2004a: 114). Unlike Gutt's definition, this one implies the imperative of using the best available substitute techniques. In contrast to Tabakowska's insistence on the equivalence of images and not only content, it simply assumes that re-painting imagery in all its dimensions may be impossible. If careful analysis proves it is so indeed, one has no choice but to be satisfied with the equivalence of content.

Although I subscribe to Hejwowski's understanding of equivalence and translatability, I borrow from Tabakowska the idea of examining the correlation of parameters of imagery. However, she deals with conventions of grammar, whereas I am concerned here with what she considers to have been investigated thoroughly and as such not worth looking at: lexical items. In my opinion, lexical items which function as terms, be it folk or scientific, behave in quite a complex fashion in translation and hence deserve attention. Terms label conceptual structures and, like grammar, can be said to be more or less conventional in a given language. Their conventionalisation involves gestalt perception and unequivocal imagery behind them. Items visualised with effort or as a batch of not quite uniform possible images are less conventional. Tabakowska analyses equivalence on the level of imagery, which is construed through conventions of grammar, mainly because this is the level of subtlety often overlooked by translators. Another reason is that lack of equivalence on this level:

> belongs among the so-called "covert errors" (House 1977) and as such, it usually involves neither simple misunderstanding of the original text by the

> translator nor the clumsiness or awkwardness of an inexpert rendering: in consequence, it is not considered as a standard "error", and as such, does not provoke the critic to compare the original with the translation. (Tabakowska 1993: 186)

In this respect, she juxtaposes non-equivalence of imagery with factual non-equivalence. In the present work it is claimed that factual non-equivalence also escapes unnoticed extremely often.

While we should be grateful to cognitive linguistics for devising a very sensitive tool for the evaluation of translation, as Tabakowska explicitly states (1993: 186), the full application of the tool incurs high costs. By 'full application' I understand at least the detailed analysis of all the dimensions of imagery as encoded in various grammatical categories and word combinations which do not directly correspond to each other in two different languages. It is then justified in the examination of translations of belles-lettres, especially poetry, where such phenomena as given/new opposition expressed via definite and indefinite articles, or the obligatory backshift of tenses may have serious consequences for the coherence, logic, imagery, originality of expression, and any other important features of literary work. In the translation of more pragmatic texts the lack of equivalence on the level of imagery rather than stemming from the violation of grammatical conventions or the objective impossibility of pairing them often results from overlooked factual non-equivalence or mismatched/ impossible to match category levels, both of which are encoded in words alone. But belles-lettres often share a significant number of characteristics with pragmatic texts. Mismatched facts/realities or categories concerning the natural world regularly pass unheeded because of the tendency of predominantly urban societies to be increasingly ignorant of concepts related to nature, particularly from levels other than basic. Not infrequently nature terminology plays as important a role in literary as in non-literary communication. For instance, it contributes to the scene construal as a setting for a scene, sometimes presaging ensuing events or giving other valuable clues. Therefore, despite the fact that the examination of imagery construction through terms rarely goes beyond looking for linguistic correlates of particular dimensions of imagery other than lexical items, it seems challenging enough. As such, it is carried out in Chapters Three and Four.

1.3.2. The linguistic view of the world framework

Another theoretical framework referred to in the present work concerns the studies on the linguistic view/image/picture of the world. Anusiewicz

(1996: 261-89) traces the origins of the concept back to Martin Luther. It was taken up in the 19th century by Wilhelm von Humboldt and more systematically developed by 20th century German linguists Leo Weisberger and Helmut Gipper. In Poland, it was introduced as late as in the 1980s by Jerzy Bartmiński and Ryszard Tokarski. Maćkiewicz (1991: 12) observes that before the concept appeared in linguistics, it had been used as 'an image of the world' in physics by Hertz and Planck. In linguistics, German studies were paralleled a little later by Edward Sapir and Benjamin Lee Whorf. The concept of a linguistic view of the world is 'an important element of the research programme of ethnolinguistics' (Bartmiński 2004: 103). Polish scholars working within this framework try to describe scientific and folk interpretations of fragments of chaotic reality encapsulated in the grammar and lexicon of a given language. They dwell on the work of Anna Wierzbicka and cognitive linguists, from Minsky, Fillmore, Taylor to Lakoff, Langacker and Fife. They present categorisation as interpretation and stress the role of the interpreter, who determines division lines (the maximal and minimal scope) of categories, the choice of prototype and their openness to changeability. Some of them aim to describe in the form of 'cognitive definition' conceptual models evoked by lexical items, which they call, after Minsky, Fillmore and Taylor, holistic interpretive frames. These are culturally interpreted views of the fragment of the world correlated with a given lexical item, which embrace the semantic core (prototype) and connotational features, both well-established and weak. Such frames consist of subframes, which for certain categories have proved to be quite regular. For example, in *Lexicography and conceptual analysis* Wierzbicka devised them for animals and artefacts (1985: 161)[8], while Tokarski in *Słownictwo jako interpretacja świata* [*Vocabulary as an interpretation of the world*] (1993: 340-341) did the same for plants. The reconstruction of the category's scope is not always possible according to a neat pattern. Researchers sometimes start not from a specific lexeme but from a chosen fragment of reality and look at words and their combinations which describe it. One such researcher is Jolanta Maćkiewicz.

The works of Maćkiewicz are particularly relevant for this analysis because she is especially interested in pre-scientific and scientific categorisation, their mutual interaction and its effects. She compares Polish and English, being preoccupied with earth sciences – an area of concern here as well. First and foremost she establishes links between

[8] Reported in Tokarski (1996: 110).

categorisation and the image of the world[9] . Thus, categorisation as a mechanism, which helps make sense of experience, allows people to move about the surrounding chaos. We obtain a simplified world, reduced to small chunks, adjusted to the limits of human cognition. Only a portion of categories is verbalised. They build a linguistic picture of the world, while the rest compose a much richer conceptual image. Verbalised categories may have – and often have – different boundaries across languages, but even if the boundaries are the same, their internal structure can vary, mainly as to the choice of the prototype. Prototypes are those entities which are cognitively most salient, most contrastive in relation to the members of other classes and which have a high frequency of occurrence. The selection of the prototype conveys the importance of a given entity in the consciousness of a particular human group, which is in turn connected with the natural environment and cultural milieu. In different languages non-prototypical exemplars can differ in how much more or less similar they are to the prototype and in the degree of similarity between each other. The differences in the general language categorisations concern the same model of reality. When folk and scientific categorisations are juxtaposed, two different models of reality emerge.

Maćkiewicz, following Bartmiński and Tokarski, identifies several types of data necessary to reconstruct the linguistic view of the world. They are:

- information included in the word itself – its etymology[10] and its meanings
- lexical derivatives of a lexeme[11]
- set phrases, proverbs, antonyms

To these she adds one more:

- the lexico-semantic field in which a word occurs (2004b: 199).

[9] It is done most succinctly in Maćkiewicz (2004a: 47-55).

[10] Etymology must be approached with caution since some words do not reflect the contemporary view of the world but preserve obsolete beliefs already verified by science and imposed on inexpert users of a given language (e.g. *wieloryb* = ‘enormous fish’ continues to be the name for ‘whale’ in Polish despite the common knowledge now that it is a mammal) – see Tokarski (1995: 55).

[11] For example, in the case of *wyspa* (island) a researcher should consider such derivatives as *półwysep* (literally, half-island = peninsula), *wyspowy* (occurring in patches), or *wyspiarski* (living on an island, characteristic of an island).

For the sake of the following study this is a very useful addition since, not aspiring to reconstruct the image of the world in its entirety and concentrating on translation problems, we can nevertheless draw helpful conclusions taking advantage of this one type of data. We focus our attention on the prototype status of a word or its position in relation to the centre of verbalised category.

The lexico-semantic field is understood here as 'a group of words bound by the common meaning between which there exist specific semantic relations aiming to form a coherent and hierarchical system'[12]. Lexico-semantic fields are inextricably linked to the corresponding conceptual fields, reflecting the way in which the world is perceived and categorised in a particular culture (cf. Marczewska 2002: 31). They are open sets whose composition depends on the arbitrary decisions of a researcher, who determines 'the common meaning' and the number of members that share it (Maćkiewicz 1991: 37). The broader the field the more difficult it is to delimit membership qualifying properties and pertinent information[13]. Inevitably, the question of categorisation surfaces again. The research material can lead scholars to arrive at surprising hierarchical structures within their selected fields. For example, Marczewska (2002: 36) discovered that Polish folk taxonomy does not classify trees as plants and that there are two parallel hierarchies: in the first, at the superordinate level, two categories appear – broadleaved[14] trees and needleleaf trees, the former taking one subordinate category of fruit/non-fruit trees, the latter having none; in the second, trees are divided into female and male, with respective subcategories presented as they go down the ladder: evil/unlucky vs. good/lucky, condemned vs. blessed, weak vs. strong. Hence the lexico-semantic field of the folk category of 'tree' comprises names of trees grouped simultaneously in two hierarchical structures, neither of them submitted to the super category of plants.

Notwithstanding the arbitrariness of the borderlines of lexico-semantic fields Maćkiewicz manages to develop criteria for their internal organisation. In other words, she tries to devise a set of aspects to be tested in order to locate the prototype of a category (the prototypical lexical item) in a given language. Thus the centre is occupied by the items which:

---

[12] Tokarski's definition quoted in Maćkiewicz (1991: 37, translation mine).

[13] See Pajewska's criteria of inclusion to the category 'woodland vocabulary' (2003: 12, 18).

[14] Although 'broadleaf' is possible, I use 'broadleaved', which seems to dominate in relevant literature.

- represent the most important differential properties
- have the widest scope of meaning
- have the least restrictive collocationality
- are stylistically and emotively neutral (foreign-sounding, dialectal, sociolectal and idiolectal elements are moved to the periphery)
- occur with the highest frequency (1991: 39).

The above observations assist in my efforts to find the prototype of corresponding categories in Polish and English texts on nature. While taking advantage of them, I am well aware of their shortcomings. For instance, Taylor considers frequency as not indicative of prototypicality. He notices that 'children's drawings depict the grass as an undifferentiated green' even though in many countries it is brownish for most of the year. He concludes 'Prototypically, grass is green, which is not the same as saying that grass is usually green' (1995: 53).

Most importantly, Maćkiewicz comes to a number of conclusions concerning scientific and non-scientific (folk, pre-scientific) categorisation, which are certainly valid for our analysis. In her dissertation *Nienaukowy i naukowy obraz morza* (1991) [*A non-scientific and scientific image of the sea*] she does it on the basis of dozens of lexico-semantic fields in Polish and English, compared with correlating terminological fields. My comparisons are not so comprehensive since, unlike Maćkiewicz's, mine is not a lexicographic study, but her method is appreciated as very revealing. The translator, who suspects that the UK, an island country, has an exuberant marine vocabulary collection can nonetheless be awestruck on discovering that in the lexico-semantic field: 'names of parts of the border between the land and the sea protruding into the sea' Polish geographers distinguish nine items, with a fuzzy boundary between two of them, while the respective English terminology includes eighteen items with a different hierarchical structure and six lexemes outside this structure. If terminologies differ so much, it is small wonder that non-scientific vocabulary in English contains as many as twelve well-defined items *vis à vis* the Polish four, of which two are interchangeable synonyms and all have vaguely staked out borderlines (cf. 1991: 89). What is most often referred to in my work is Maćkiewicz's findings (mainly 1991 and 1996) concerning the mutual relations between scientific and non-scientific models of linguistic and extralinguistic reality. She observes that[15]:

1. non-scientific models
   - look for analogies;
   - prefer categories based on contrast, value judgements and metaphoric extensions to hierarchical organisation;
   - rarely use taxonomic classifications relying on typologies (categories are broad, not necessarily internally uniform);
   - have incomplete hierarchical structures (e.g. no hypernym, only one hyponym);
   - are asymmetric (relations between hyponyms are unclear, criteria for organisation are numerous and changeable);
   - are constructed from the anthropocentric, ethnocentric and egocentric perspective (personification of natural phenomena, enhancing their utilitarian qualities, noticing primarily properties perceived through eyes such as size or shape, limiting oneself to the familiar and the close);
   - are polythetic – based on many shared features.

2. scientific models
   - look for differences;
   - prefer hierarchical structuring, with lower levels maximally divided;
   - replace types with classes, internally uniform and with well-defined boundaries;
   - have hierarchies designed in such a way so as to be able to accommodate new data;
   - strive to preserve symmetry – co-hyponyms are juxtaposed on the basis of precise criteria;
   - use the bird's eye view perspective aiming to cover the widest range possible;
   - in some disciplines eliminate the utilitarian factor;
   - aspire to be monothetic (based on one common feature or several properties forming one whole).

However, naïve and expert models have always influenced each other and therefore scientific categories not always have closed and clear-cut

[15] These observations are compiled on the basis of all Maćkiewicz's works cited, especially (1991) and (1996).

boundaries. They take into account properties visible to the naked eye and not only those testable objectively. All this leads to the adoption of general language words by terminology, the terminological re-naming of prescientific concepts, the co-existence of proper names reflecting the inexpert perspective and terms which contrast in meaning (The Bay of Biscay, for example, is not a bay but an adjacent sea). On the other hand, non-scientific language adopts certain characteristics of expert terminology – the broadened perspective (names of objects which do not occur in the neighbourhood, such as fjord and taiga in the case of Londoners, for example, become used and understood), folk categories become more precisely defined and their scope is narrowed down. Last but not least, expert models are numerous. Unavoidably then, the translator is forced to face interterminological polysemy. When it concerns fairly distant disciplines the probability of misunderstanding is low – for instance, the Polish *zatoka* co-occurs in geography ('bay'), anatomy ('sinus') and meteorology ('depression'). But in the case of closely related disciplines there is never enough caution, which Chapter Two of this book highlights and richly exemplifies.

Reaching beyond one-language confines, the translator as the interpreter of an interpreter re-categorises the reality in the target language and should not let chaos back into what has already been arranged in a particular way. The translator must carefully re-create the reality behind concrete terms or, in other words, the view of the world deposited in them. Unlike most ethnolinguists, I do not try to offer detailed and comprehensive descriptions of a view of a fragment of the world, but Chapters Two, Three and Four will present what interpretation or image a term projects in a context. The translatability of categories appears to require lengthy elaboration even if we refrain from the ambitious task of reconstructing all possible aspects of an interpretive frame. A convenient illustration of it is accidentally included in one of the most prominent Polish books on the linguistic view of the world: *Językowy obraz świata* [*A linguistic view of the world*] (Bartmiński (ed.) 2004). Bartmiński notices there that words are defined in relation to a viewpoint which determines the set of interpretive subframes, and proceeds to enumerate those sets for the word *bławatek* (*Centaurea cyanus*, cornflower) from the point of view of the farmer and the tourist (2004: 103-120). The article is written in Polish but the book also includes an English summary, in which *bławatek* is rendered as 'bluebottle'. Such a choice of equivalent gives rise to questions of a different nature: why the non-canonical 'bluebottle' instead of the currently recommended 'cornflower'?; or more importantly: why was the more ambiguous option selected? Bluebottle also refers to any blow-fly or to its one representative, the common kitchen nuisance blue bottle fly (*Calliphora vomitoria*), and finally – which referent, for the fly or for the flower is in more widespread usage? In other words, whose viewpoints should the translator take into account? Simply not to get sidetracked, the question about the vagaries of categorisation that, amusingly, bring flies and weeds together, or about all sorts of weak connotations of the word 'bluebottle' or 'cornflower' as well as *bławatek*, however intriguing, must be omitted. A similarly relevant situation is found in the same volume, in R. Grzegorczykowa's article (ibid.: 39-46), where, having defined the linguistic image of the world, she enumerates what else it also includes, and such association derivatives as *papierówka* are mentioned. It would indeed be enticing to explicate why this apple cultivar is associated with paper in Polish and why it sounds so very homely but, for the translator, the most pressing issue is whether 'pearmain' given as the equivalent of *papierówka* in the English summary is in fact the same thing and, on assumption that it is (although it is doubtful), whether 'pearmain' is a good instance of association derivative under discussion in the English version of the text. In the context, the translator – rather than tracing the images behind both words – has to concentrate on finding a pertinent example of the derivative. In general, the idea of the linguistic image of the world is borrowed to underline the role of equivalence on the level of imagery. However, the methodology worked out by scholars reconstructing this image is treated very selectively to meet the needs of Translation Studies.

CHAPTER TWO

THE CATEGORISATION OF NATURE IN POLISH AND ENGLISH SPECIALIST TEXTS

2.1. Categorisation versus translation

In this chapter the intricacies of the categorisation of the natural world by specialists are elaborated, with talk of 'specialist' categorisation to cover a wider range than suggested by the modifier 'scientific' preferred by Maćkiewicz. It meshes with the nomenclature of terminological studies and tallies with the adjective 'expert' used by Taylor (1995). Specialists and experts are not always scientists. More often than not they are practitioners with relevant scientific or scholarly training. The texts analysed are usually written by such practitioners and this has another important consequence: they do not represent a uniform scientific perspective. The linguistic categories the texts contain are properly understood via their background framing – the author's motivation, goals, speciality, place of origin or even teachers he or she follows. A great deal of space below is devoted to the interdisciplinary polysemy of terms and endeavours to keep it in check, and, more ambitiously, to minimise it. Some attention is drawn to the influence folk categorisation has on specialist classification in Polish and English. For instance, applied disciplines, like forestry, tend to categorise on the basis of utilitarian values. As only selected properties are of interest to somebody who does the classification, natural categories allocated a secondary role have fuzzy boundaries – foresters use the folk term *bagno* in their documentation, gathering in it habitats spanning all damp sites, from rare and protected to common and degraded beyond repair. Simultaneously, two other cognitively valid features of categorisation are explored: stretchability and gradability. Conservationists move boundaries of categories and accept new members, excluding others, to accommodate a rich diversity in a compromised system. Such systems, of which Natura 2000[16] is pivotal in the European Union, presuppose the flexibility of categories and consequently the changeability of their prototypes – across regional variations and according to agreements between advocates of unrestrained exploitation of nature and those opting for sustainability.

Scientists who embark on classifying components and phenomena of the world must acknowledge that 'there may be as many classifications of any series of natural, or other, bodies, as they have properties or relations to one another, or to other things; or, again, as there are modes in which they may be regarded by the mind …' (Huxley cited in Joosten 1998, unpaginated). Classification is then an effort to arrange these properties and relations so as to reduce difference and maximise useable information. And as Joosten observes: 'Information is no objective feature but depends on existing knowledge, [which] determines the resolution of observed difference' (1998, unpaginated). Knowledge progresses with time. At a given time it also varies across scientific disciplines and countries due to limitations in access to data and the motivation behind a particular research. If the motivation is the conservation of natural diversity, it stands at odds with the fundamentals of classification. Diversity conservation implies noticing and enhancing as much difference as possible. Classification recognises that 'things (or concepts) have likeness' (ibidem). The fewer categories a classification has the more easily remembered it becomes. If a classification is to serve diversity conservation it has to strike a balance between useability and safeguarding a sufficient quantity of identifiable 'natural bodies'. Useability means enabling fast identification, smooth communication and accommodation of new information as well as allowing prediction and interpolation. The world evolves and so do classifications. As in nature, this evolution is not a coordinated process. Despite attempts, unification of the products of classification – taxonomies, nomenclature and related terminology – has not materialised yet, and stands little chance of doing so. These products keep developing independently.

Labels proliferate, which engenders the necessity of intralingual translation between various scientific disciplines. In cognitive terms, intralingual translation results from essential features of categories:

---

[16] The European Union network of sites designated by Member States under the Birds Directive of 1979 (called Special Protection Areas) and under the Habitats Directive of 1992 (called Sites of Community Importance, of which the European Commission chooses and designates Special Areas of Conservation), to contribute towards ensuring biodiversity through the conservation of natural habitats and of wild fauna and flora.

a) background framing – one needs to translate categories individualised from one viewpoint into those singled out from another;
b) fuzziness – one must look for borderlines of categories in the two compared interpretations of the world to see to what extent they cover common ground;
c) gradability – one must recognise which peripheral members of a category remain so in corresponding categories in a compared interpretation, and which of them move closer to the core or leave the set;
d) stretchability – one must spot which members of one category are moved to another to fit the criteria adopted by the two compared interpreters, and which of them appear in more than one category in both interpretive systems.

As for interlingual translation, in order to ensure that a translated text uses appropriate terms and names, it requires careful consideration of multiple aspects – such as credibility of available dictionaries and relevant scientific sources, or suitability of nomenclature for the purposes of a given publication. Again, the background framing and the demarcation lines of verbalised categories must be compared before decisions on the application of particular terms are taken. In the present chapter all the issues signalled above are addressed, from the difficulties of intralingual and interlingual translation for the Natura 2000 programme, through endeavours to enable universal translatability of mire terminology, to the discrepancies between Polish and English nomenclature of more complex categories, such as habitats, and more basic ones, such as species.

Classifications of natural habitats can serve roughly two purposes: exploitation and conservation. In order to protect nature successfully, it is necessary to protect entire ecosystems, not just individual species. The more precise the description of an ecosystem is, the more effective its conservation (cf. Pawlaczyk and Mróz 2003: 58). Precision is one requirement, agreement between conservationists and landowners – state or private – whose main interest is production is another. Agreement in the sense of compromised plans in reference to particular fragments of woodland, peatland or grassland must be preceded by establishing a mutually translatable typology of ecosystems. The most pertinent example is the classification of ecosystems used by foresters and the one needed for the Natura 2000 programme. Before discussing the problems and possibilities of this intralingual translation it should be noted that even the very meaning of 'natural habitat' and 'habitat' must first be disambiguated.

Pawlaczyk and Mróz (ibidem) explain that the Habitat(s)[17] Directive term 'natural habitat' equals 'ecosystem' in meaning, which, to simplify, is biocenosis (a natural set of living organisms in a state of dynamic balance) plus biotope (an area clearly distinguished from others by a specific combination of abiotic conditions). The Polish equivalent must then be *siedlisko przyrodnicze*. Unfortunately, a different misleading translation has penetrated into official documents – *siedlisko naturalne*. In Polish *naturalny* signifies 'unchanged by human activity' or 'one that would develop if not for human intervention'. The point is that the Habitats Directve aims to preserve a wide range of habitats, including those that owe their existence and valuable biodiversity to extensive human management. Hay or litter (*Molinion*) meadows are worth as much as the intact deciduous forests of Białowieża. The adjective that covers both these mutually exclusive meanings is *przyrodniczy*. Interestingly, in English, no such distinction is made and 'natural' can stand for *naturalny* or *przyrodniczy,* depending on the context.

To make matters more complicated, the notion of habitat (*siedlisko*) as used by Polish foresters does not fully overlap with the notion of natural habitat. It is defined as a complex of abiotic factors: climatic and soil conditions on a specific geographic and topographic location (based on Szymański 2000: 11-14). That has far-reaching consequences, since for forestry the conservation of a habitat relies on preventing soil from erosion and from the unfavourable impact that tree stands of inappropriate composition exert on this soil, and on maintaining suitable soil-forming processes. In order to protect a natural habitat one has to take care of all the above, and also of its structure and floral composition together with the associated fauna and ecological processes. In effect, much more complex knowledge and work is required and searching for compromised solutions becomes a necessity: growing best-quality timber often conflicts with maintaining biodiversity. The content of the category 'habitat conservation' is so dependent on background framing that it can undergo drastic alterations. Given that in conservationist jargon *siedlisko przyrodnicze* is often referred to simply as *siedlisko,* misunderstandings are no surprise. In addition to all that, there exists a term 'species habitat', a site where a particular species finds conditions favourable enough to live in. These can occur across a number of natural or semi-natural habitats.

---

[17] The forms 'Habitats' and 'Habitat' are both used equally frequently but in this work 'Habitats' will be used throughout.

2.2. The intra- and interlingual translation of woodland habitats

In an endeavour to create a common European classification of nature and come up with a common language for European naturalists, the international CORINE (Coordination of Information on the Environment) programme was launched in 1985. As a result, the PHYSIS coding system classifying Palearctic habitats was devised in 1996 and adopted in the *Interpretation manual of EU habitats* published by the European Commission for the needs of Natura 2000, and updated after the accession of new members. As claimed in Pawlaczyk and Mróz (2003: 59), other classifications, such as the phytosociological classification of plant assemblages popular in Central Europe, are hardly known elsewhere. A forestry typology of woodland for management purposes varies from country to country, and is 'untranslatable'. The Polish forestry's typology, based on the work of Mroczkiewicz and Trampler of 1964 (Szymański 2000: 385-98), is translated as follows (Fronczak 2003: 150):

Bb (*bór bagienny*) – marshy coniferous forest
BbG (*bór bagienny górski*) – montane marshy coniferous forest
BG (*bór górski*) – montane coniferous forest
BMb (*bór mieszany bagienny*) – marshy mixed/coniferous forest
BMG (*bór mieszany górski*) – montane mixed/coniferous forest
BMś (*bór mieszany świeży*) – fresh mixed/coniferous forest
BMw (*bór mieszany wilgotny*) – wet mixed/coniferous forest
BMwyż (*bór mieszany wyżynny*) – upland mixed/coniferous forest
Bs (*bór suchy*) – dry coniferous forest
Bśw (*bór świeży*) – fresh coniferous forest
Bw (*bór wilgotny*) – wet coniferous forest
BWG (*bór wysokogórski*) – high-mountain forest
LG (*las górski*) – broadleaved montane forest
Lł (*las łęgowy*) – riparian forest
LMb (*las mieszany bagienny*) – marshy mixed/broadleaved forest
LMG (*las mieszany górski*) – montane mixed/broadleaved forest
LMśw (*las mieszany świeży*) – fresh mixed/broadleaved forest
LMw (*las mieszany wilgotny*) – wet mixed/broadleaved forest
LMwyż (*las mieszany wyżynny*) – upland mixed/broadleaved forest
Lśw (*las świeży*) – fresh broadleaved forest
Lw (*las wilgotny*) – wet broadleaved forest
Lwyż (*las wyżynny*) – upland broadleaved forest
Ol (*ols typowy*) – alder woodland
OlJ (*ols jesionowy*) – alder-ash woodland

The above renderings are widely used in other State Forests publications[18], sometimes with slight changes ('wet' is replaced by 'moist')[19]. Consistency vanishes in publications by external institutions. Some regional directorates of State Forests have obtained SmartWood certification[20], the reports on which are published in English. In one of them, prepared for the Radom and Łódź regions, the following nomenclature examples can be observed:

- "*grądy*" – mixed deciduous oak-hornbeam forests with maple, linden, or others (the Polish word, followed by a hyphen, is part of the English name used)
- termophilous oak forests
- alder alluvial forests
- alder carrs
- bog pine or birch forests
- fresh pine forests
- alluvial alder-ash forests and alder forests

(Smartwood 2004: 11-12)

The names used depart from official forestry typology and look like an attempt to translate Polish names of woodland habitats used by conservationists with an occasional and not very 'faithful' borrowing from the forestry nomenclature, such as 'fresh pine forest'.

An attempt to translate forestry typology into the language of nature conservation more systematically and comprehensively is undertaken by Pawlaczyk and Mróz (2003). They translated this typology into the language of Habitats Directive, with the view to securing protection to those fragments of Polish state forests that meet Natura 2000 criteria. The authors have since added many details by publishing comparative studies so as to provide guidance to foresters and environmentalists[21]. In early

---

[18] Although Fronczak (2003) does not give the translator's name, James Richards, who translated many other State Forests publications, seems most likely.

[19] See e.g. Zajączkowski (1999).

[20] It is a forest management certification programme launched in 1989 by the Rainforest Alliance, a US-based international organisation cooperating with the Forest Stewardship Council (FSC) – an independent, non-governmental, not for profit organisation, established to promote the responsible management of the world's forests.

[21] Pawlaczyk and Mróz (2003), http://www.lkp.org.pl/poradniki/lasy/fotoprzewodnik_lesne.pdf and http://www.lkp.org.pl/poradniki/lasy/fotoprzewodnik_nielesne.pdf

2007 their efforts brought fruit in the shape of the official ordinance by the director of the State Forests instructing foresters how to survey and register woodland and non-woodland habitats. Also, *Poradniki ochrony siedlisk i gatunków* [Guides on habitats conservation and species protection] have appeared on the official Natura 2000 website of the Ministry of Environment, which define habitats, determine their subtypes within broad Habitats Directive categories and relate them to the Polish phytosociological classification and sometimes the Polish forestry typology of woodland. On the ensuing pages, with the help of examples from Pawlaczyk and Mróz (2003), only a fraction of habitat translation problems are presented.

While correlating phytosociological classifications and Natura 2000 natural habitats is not very difficult, in practice most State Forests areas do not have maps and descriptions of actual plant communities. They only offer maps of woodland types, with which natural habitats correlate in an extremely complex fashion. Admittingly, woodland types allow a natural scientist to make certain presumptions – which must then be verified on the spot. For instance, a site marked Bb can be expected to be the bog woodland (code 91D0)[22], whereas OlJ – alluvial forests with *Alnus glutinosa* and *Fraxinus excelsior* (code 91E0). But most often a site with one woodland type may hold several natural habitats, and one natural habitat may be found in many woodland types. Moreover, this also depends on the geographical location – Lśw in Pomerania usually hosts acidophilous beech forest (*Luzulo-Fagetum* beech forests – code 9110), in central Poland oak-hornbeam forests (Sub-Atlantic and medio-European oak or oak-hornbeam forests of the *Carpinion betuli* – code 9160), and in Małopolska thermophilous oak woods (Euro-Siberian steppic woods with Quercus spp. – code 91I0) can appear in patches.

The data available from forestry inventories concerning the herb layer, indispensable to determine a natural habitat, are incomplete because as a rule only five most common species have been recorded. That leads to situations where a beech stand in Pomerania, marked Lśw, might turn out to be an oak-hornbeam forest (*grąd*), not beech forest; an alder stand, marked Ol, may be an alder swamp forest (*ols*), which is not protected within the Natura 2000 programme, or an alluvial forest, code 91E0; in turn, an oak stand in LMśw, may in western Poland be old acidophilous oak wood with *Quercus robur* on sandy plains – code 9190 or an oak-hornbeam forest, or thermophilous oak wood (Euro-Siberian steppic woods with Quercus spp. – code 91I0). Some natural habitats tend to

[22] The Natura 2000 codes and names corresponding to them are quoted after the *Interpretation manual of EU habitats.*

correspond to one woodland type, for instance, *górskie jaworzyny ziołoroślowe* (a subtype of Habitats Directive Medio-European subalpine beech woods with *Acer* and *Rumex arifolius*, code 9140, labelled in the EUNIS[23] system 'tall herb mixed sycamore forest', EUNIS code G1.A413) occur in LGśw sites; others coincide with two or three (e.g. *bory chrobotkowe* – Central European lichen pine forests, code 91T0) in Bs and Bśw; but the majority can be found in several types, for example, *ciepłolubne buczyny storczykowe* (Medio-European limestone beech forests of the Cephalanthero-Fagion – code 9150) in LMśw, Lśw, LMwyżśw, Lwyżśw and LGśw. Annex II to *Decyzja nr 5 Dyrektora Generalnego Lasów Państwowych z dnia 30. 01. 2007 w sprawie metodyki inwentaryzacji siedlisk i roślin* (henceforth *Decyzja nr 5 30. 01. 2007*), [The decision no. 5 of the Director General of the State Forests of 30 January 2007 on the methodology of the habitat and plant surveys], in some cases, gives even more possibilities. For instance, *grąd środkowoeuropejski lub subkontynentalny* (or simply, *grąd typowy*) – *Galio-Carpinetum* oak-hornbeam forests (code 9170) – should be looked for in ten woodland types. From the perspective of woodland types Lśw is a record holder when it comes to the number of Natura 2000 habitats that it potentially embraces: oak-hornbeam forests (9160, 9170), beech forests (9110, 9130, 9150), termophilous oak forests (91I0), and slope forests (9180) – seven types altogether. The stretchability of category boundaries peaks here.

From a practical point of view, translating forestry documentation into conservationist language requires detailed field studies with emphasis on the identification of herb layer species and assessment of the level of naturalness and degradation of a given site. From the point of view of a nature-loving tourist, who is equipped with a very generalised guidebook or interpretation board information and looks superficially at the current tree composition, some names may seem inadequate and puzzling. For example, in Annex I to *Decyzja nr 5 30. 01. 2007* the reader is informed that *kwaśne buczyny* – acidophilous beech forests (habitat 9110) are sometimes montane fir forests, even without the beech; *górskie jaworzyny*

[23] The EUNIS (European University Information Systems) data are collected and maintained by the European Topic Centre on Biological Diversity for the European Environment Agency and the European Environmental Information Observation Network to be used for environmental reporting and for assistance to the NATURA 2000 process (EU Birds and Habitats Directives) and coordinated to the related EMERALD Network of the Bern Convention. I cite the EUNIS habitat type names if the *Interpretation manual of EU habitats* does not offer any credible equivalent of a Polish habitat.

ziołoroślowe (habitat 9140) might happen to be beech forests, even though *jaworzyna* suggests the dominance of the sycamore. In turn, Annex II warns that in upland areas *grąd* (oak-hornbeam forest: code 9170 habitat) with the fir and spruce can be mistakenly classified as *jedlina* (fir forest: 91P0) or *buczyna* (beech forest: 9110, 9130). In an Internet guide[24] one can see a photograph of *żyzna buczyna górska* (habitat *Asperulo-Fagetum* beech forests – code 9130), where most of the space is occupied by spruce trees with sparse beeches and firs. What counts, though, is the field layer composition, characteristic of beech forests. If the forest floor does not support any such ground flora, it indicates that the habitat is degraded enough to bear the name based on its superficial appearance – *sztuczna świerczyna* (spruce plantation).

Still more surprisingly to a layman, Annex II of *Decyzja nr 5 30.01.2007* sensitises foresters and ecologists to the fact that forestry documents (mainly maps) may present a particular forest type (and natural habitat at the same time) – *łęg* (code 91E0) as *jezioro* (a lake). This refers to small patches of alluvial forests clumped together with the surrounding aquatic habitat by a map maker. No monolingual dictionary defines 'lake' as 'riparian forest', 'riverine forest' 'ash-alder forest/carr', 'alluvial forest', or 'floodplain forest'. Neither does any bilingual Polish-English dictionary advise translators to occasionally render *jezioro* as one of the six English equivalents of *łęg* listed above. The forestry categories of *łęg* and *jezioro* are fuzzy and they overlap in a way that provides an apt illustration of the stretchability of categories.

The registration of a habitat under one name and not another, with which it can be easily mixed up, has often serious consequences. The *ols/łęg* opposition implies a choice: grant a given site protection or not. *Ols* – alder swamp forest is not on the Habitats Directive list, while *łęg* – alluvial forest, although often superficially a look-alike, is. Here, it is a matter of survival. In the example to follow, both habitats enjoy a priority conservation status: oak-hornbeam forest (*grąd*) and termophilous oak forest (*dąbrowa świetlista*). But classifying the latter as the former causes the slow death of *dąbrowa świetlista*. This habitat requires active conservation, as leaving it to its own devices will certainly allow natural succession towards oak-hornbeam forest to take place. In the past such open oak stands were created by grazing livestock in them. Animals stemmed the growth of shrubs and enabled small termophilous plants to flourish. In our climate the cessation of grazing brings about an immediate invasion of shrubs and deciduous tree saplings, which soon shade and

---

[24] At http://www.lkp.org.pl/poradniki/lasy/fotoprzewodnik_lesne.pdf

choke light-loving herb layer vegetation. In dry climates this process is much slower, and even wild animals in their small numbers manage to maintain the suitable conditions for low xerothermic flora. To sum up, correct classification is only the first step. Without a practical implementation of conservation recommendations based on studies necessary for classification purposes, the rarest, niche habitats will always suffer. But without the first step no other is possible. Thus, in the case of natural habitats the question 'what's in a name?' can be answered 'either destruction or prolonged existence'.

As shown above, intralingual translation for the needs of nature conservation is a complex matter embedded in categorisation, which cognitivism teaches us to be prototype-based. Interlingual translation is even more serious. Although the names of basic Natura 2000 habitats were translated into Polish in Dyduch-Falniowska et al. (2002), subtypes differentiated there as an adjustment to Polish conditions, were not, and remain so several years later. In the meantime *Poradniki ochrony siedlisk i gatunków*[25] have refined some Polish names, which, however desirable from a scientific point of view, add to the longish list of habitat names in circulation. They have not got an English version, either. Disappointingly, the *Interpretation manual of EU habitats,* updated after Poland's accession to the EU, has often limited itself to comments and descriptions rather than offering convenient labels. Although it correlates the British National Vegetation Classification (NVC), and the German and Nordic classifications, with the corresponding Habitats Directive names, the Polish phytosociological classification is not included. It comprises references to PHYSIS codes, which enable the successful search of an English name as long as it belongs to a separate entry. If it does not, one can try looking it up in the EUNIS classification, like the previous one, available at www.nbn.org.uk /habitats/-11k (National Biodiversity Network (NBN) Habitats Dictionary). For a non-specialist this is extremely time-consuming since only occasionally correlations with other classifications are provided, and one is left with brief phytosociological notes as clues.

To exemplify the complexity of the task of correlating various labels for natural habitats, excerpts of the four sources mentioned in the previous paragraph are analysed below. For convenience, Dyduch-Falniowska et al. (2002) is referred to as **A**, *Poradniki ochrony siedlisk i gatunków* as **B**, PHYSIS database as **C**, and EUNIS database as **D**. For a start, let us compare the Natura 2000 habitat 91E0. **A** labels it as *łęgi i górskie olszyny*, **B** as *łęgi wierzbowe, topolowe, olszowe i jesionowe, olsy*

[25] Available at http://natura2000.mos.gov.pl/natura2000/pl/poradnik.php - 86k

źródliskowe. Both are an equivalent of 'Alluvial forests with *Alnus glutinosa* and *Fraxinus excelsior* (*Alno-Padion, Alnion incanae, Salicion albae*)'. **A** distinguishes the following four subtypes: *łęg jesionowo-olszowy, podgórski łęg jesionowy na stanowiskach niżowych, łęg topolowo-wierzbowy and nadrzeczne i nadpotokowe olszyny górskie*. As a result of various negotiations the official version represented by **B** differentiates seven subtypes: *łęg wierzbowy, łęg topolowy, niżowy łęg jesionowo-olszowy, źródliskowe lasy olszowe na niżu, podgórski łęg jesionowy, nadrzeczna olszyna górska* and *bagienna olszyna górska*. There is no point dealing with **A**'s correspondence with English versions because it was superseded by **B**. Then, in **C,** the following can be correlated: *łęg wierzbowy* and *łęg topolowy* – white willow gallery forest, *niżowy łęg jesionowo-olszowy* – medio-European stream ash-alder woods, *podgórski łęg jesionowy* – sedge ash-alder woods (probably but not certainly), *nadrzeczna olszyna górska* – montane grey alder galleries, *bagienna olszyna górska* could be montane grey alder galleries, but may partly belong to meso-eutrophic swamp alder woods, and *źródliskowe lasy olszowe na niżu* – ash-alder woods of rivulets and springs (probably but not certainly). **D** provides us with the information that mixed riparian floodplain and gallery woodland, current code G1.2, overlaps with Alluvial forests with *Alnus glutinosa* and *Fraxinus excelsior* (*Alno-Padion*, *Alnion incanae*, *Salicion albae*), code 91E0 and so does *Fraxinus-Alnus* woods of rivulets and springs, code G1.211. Besides, there is a list of subtypes that are not said to be related to Habitats Directive habitats, but for a non-specialist some of them, about eleven, cannot be decisively excluded, if they can occur in Central Europe and bear such names as G1.2111 sedge ash-alder woods, *Fraxinus-Alnus* woods of fast-flowing rivers G1.212, or Central European slow river ash-alder woods G1.21311. It seems that if specialists have not matched the existing labels or proposed their English form for five years, laymen stand little chance of doing this on their own properly.

Another illustration shows how easily a non-specialist can make an error, mainly due to a clash between the scientific search for differences at lower taxonomic levels and the tendency of folk categorisations to seek similarity regardless of the level (cf. Maćkiewicz 1991: 142-143). Within code 9170 **A** gives *grąd środkowoeuropejski* and adds 917P – *grąd subkontynentalny*, **B** distinguishes three subtypes without proposing an extra code number: *grąd środkowoeuropejski, grąd subkontynentalny* and *grądy zboczowe* (*Acer platanoides – Tilia cordata*). All three are grouped under the same PHYSIS code 41.2 oak-hornbeam (Quercus-Carpinus betulus) forests. The source does not inform us, though, that the subcodes

41.261 wood bedstraw oak-hornbeam forest and 41.262 mixed lime-oak-hornbeam forest are actually good translations of the first two Polish names respectively. Luckily, **D**'s G1.A161 and G1.A162 are identical to 41.261 and 41.262, hence the work of the translator becomes smoother. It is the third subtype that sets a trap, though. It is temptingly similar to the name of 9180 habitat: *Tilio-Acerion* forests of slopes, screes and ravines. Both lime (*Tilia*) and maple (*Acer*) are mentioned in the context of slopes (*zbocza*), so one might rush to conclude that the search for the third subtype's name is over. A closer look at the explanations under code 9180 helps reject such hasty conclusions. The maple here is *Acer pseudoplatanus* – the sycamore maple and the Polish name of 9180 habitat erases whatever hope has been left: *jaworzyny na stokach i zboczach* – nothing about the Norway maple (*klon zwyczajny*). Instead, we have *jawor,* which forms distinctive habitats. Appearances are deceptive but carefulness and diligence leaves the non-specialist translator none the wiser as to the name for the third subtype of oak-hornbeam (*Quercus-Carpinus betulus*) forests.

The last set of examples concerns habitat 91D0 bog woodland – *bory i lasy bagienne.* At first glance everything should be easy because already the *Interpretation manual* lists four subtypes, which, as the codes are provided, can be quickly located in **C**. On inspection, it turns out that, in fact, there are differences. The manual's Scots pine bog woods become Scots pine mire woods, whereas sphagnum spruce woods – mire spruce woods. A comparison with the Polish sources reveals further discrepancies. **A** differentiates five types, not four: *brzezina bagienna, las brzozowo-sosnowy, sosnowy bór bagienny, górskie torfowiska wysokie z sosną błotną* and *borealna świerczyna na torfie.* Thus, one habitat, the second in the list, has no English name to be matched with. **B** seems to exacerbate the situation by singling out six subcategories and altering three names: *brzezina bagienna, sosnowy bór bagienny, górskie torfowiska wysokie z sosną drzewokosą i kosodrzewiną, podmokła i torfowiskowa świerczyna górska, borealna świerczyna bagienna* and *sosnowo- brzozowy las bagienny.* Despite an increased number, the problem is still with only one and the same subtype, however different their labels: *las brzozowo-sosnowy alias sosnowo-brzozowy las bagienny* because the others can be linked to the subtypes **C** contains. For the sake of clarity they are presented below:

brzezina bagienna – sphagnum birch wood
sosnowy bór bagienny – Scots pine bog wood

górskie torfowiska wysokie z sosną błotną / górskie torfowiska wysokie z sosną drzewokosą i kosodrzewiną – mountain pine bog woods (following indicated PHYSIS codes but it can potentially be interpreted as a form of raised bog)
borealna świerczyna na torfie / borealna świerczyna bagienna – sphagnum/mire spruce wood
podmokła i torfowiskowa świerczyna górska – mountain sphagnum spruce wood
sosnowo-brzozowy las bagienny/ las brzozowo-sosnowy – ?

For a non-specialist correlating **C** and **D** is a real challenge: *brzezina bagienna* is probably G1.51 Sphagnum *Betula* woods. The multitude of likely candidates to denote the rest precludes responsible selection.

It could be expected that since the Habitats Directive is written in English and the National Vegetation Classification (NVC) codes are placed under its relevant codes in the *Interpretation manual*, the British conservationists do not have to do translation exercises similar to their Polish counterparts. However, as demonstrated by Hall and Kirby (1998), a lot of comparative studies of various classifications were needed in the UK to make the Biodiversity Action Plan (BAP)[26] work. The NBN Habitats Dictionary lists nineteen habitat classifications in English, some of them pan-European, others restricted to the UK, some of them already obsolete but certainly alive in older publications, widely used by non-specialist nature enthusiasts. Hall and Kirby's remark proves that they agree in opinion with Pawlaczyk and Mróz (2003): 'The names [of woodland types] used must, therefore, be treated as convenient labels, rather than as definitions' (1998: 6). Before, on the same page, they give examples in support of this claim: ' "upland woods" can exist at low altitudes, "western oak woods" can occur in the east, "oakwoods" may, in places, contain very little oak – [as when] dominated by birch and rowan during the regeneration phase'. The authors discuss nine BAP Priority Habitats in relation to the types differentiated in five other classifications: National Vegetation Classification (NVC) devised by Rodwell, Stand Type Classification, Forestry Authority Practice Guide typology, CORINE biotope classification and Annex I of the Habitats Directive. The authors observe that 'most of the [BAP] Priority types can be easily described in

[26] The Biodiversity Action Plan (BAP) is an internationally recognised programme designed to protect and restore threatened species and habitats. The original impetus for this plan derives from the 1992 Convention on Biological Diversity. The countries which have ratified this convention (hitherto 191) should develop their own BAPs.

terms of one or more NVC communities, which is not always possible using the other classifications.' (ibid.: 12). Of particular interest to us are four BAP types because, to varying degrees, they cover habitats found also in Poland: wet woodland, native pine woodlands, lowland oak and mixed deciduous woodland and, least of all, upland mixed ash woodland.

'Lowland oak and mixed deciduous woodland' overlaps with the largest number of habitats present in Central Europe, which is best illustrated by the main Habitats Directive types it corresponds to, namely: 9160 which now is Sub-Atlantic and Medio-European oak or oak-hornbeam forests of the *Carpinion Betuli*, but Hall and Kirby give it as 'Stellario-Carpinetum oak hornbeam forests' – *grąd subatlantycki*, 9180 *Tilio-Acerion* forests of slopes, screes and ravines – *jaworzyny na stokach i zboczach* and 9190 Old acidophilous oak woods with Quercus robur on sandy plains – *dąbrowy acidofilne i acidofilny las brzozowo-dębowy*. Five CORINE biotopes match those three. The number of Forestry Authority Practice Guide types is the smallest: 1) lowland acid beech and 2) oak woods and lowland mixed broadleaved woods. But Stand Type Classification, with sixteen, and NVC, with three communities but jointly nine subcommunities, beat the others. Despite difficulties in intralingual translation experienced by nature conservation practitioners, English names pose no problem when it comes to rendering them in Polish. Unlike Polish labels, which are very short and convenient but often not indicative of an immediately obvious referent, the English ones can be translated literally. To illustrate, *grąd* has several English equivalents depending on its tree composition, *łęg*, as already shown on page 34, has six, regardless of its tree composition and many more if its various combinations are taken into account. The English nomenclature is based on dominant tree species, wetness, location, and at a more specific level, on main ground layer species. This is true about some of its Polish counterparts: forestry typology of woodland and phytosociological classification at the level of plant assemblages. Hence, it is perfectly acceptable to render wet ash-maple woods as *wilgotne lasy jesionowo-klonowe* or English pedunculate oak – birch – wavy hair-grass woods as *angielskie lasy z dębem szypułkowym, brzozą i śmiałkiem pogiętym.*

Another broad BAP type worth consideration is wet woodland. It coincides with two main Habitats Directive types: bog woodland and residual alluvial forests, changed afterwards into: alluvial forests with *Alnus glutinosa* and *Fraxinus excelsior* (*Alno-Padion, Alnion incanae, Salicion albae*), four CORINE biotopes, seven types plus two subtypes according to NVC, six Stand types and just one Forestry Authority Practice Guide type. For the first two, the Polish scientists have officially approved labels: *bory i lasy bagienne* and *łęgi wierzbowe, topolowe, olszowe i jesionowe, olsy źródliskowe*. The four CORINE types pose a certain difficulty, though. While white willow gallery forests, as shown above on page 36, are *łęgi wierzbowe*, mire willow scrubs, birch and conifer swamp woods and ash-alder woods of rivulets and springs have no established equivalents. Translators may then be forced to do their best, drawing on the existing Polish names of descriptively corresponding habitats. Thus, a possible outcome may be:

- mire willow scrubs – *bagienne zarośla wierzbowe*
- birch and conifer swamp woods – *brzeziny i bory bagienne*
- ash-alder woods of rivulets and springs – *lasy łęgowe bagienne i wilgotne* (to use a not very common but practical division proposed by Sokołowski, Kliczkowska, and Grzyb (1997), which seems to come in handy here).

In contrast, the remaining labels can be easily rendered literally, for instance, slope alderwoods as *olszyny zboczowe* or bird cherry alderwoods as *olszyna czeremchowa*, or *olszyna z czeremchą zwyczajną*.

As already signalled, translating habitat names is not always a tiresome task, for which a great deal of caution is demanded. That is largely true about translating general forestry classifications based on most conspicuous floristic features of habitat physiognomy, both from Polish into English and vice versa. The same holds at lower levels of phytosociological classification, but only to some extent. As illustrations earlier in the text come from English, a handful of Polish labels are given below:

A. Forestry classification
las mieszany świeży – fresh mixed/broadleaved forest
las mieszany wilgotny – wet mixed/broadleaved forest
las mieszany wyżynny – upland mixed/broadleaved forest

B. Lower levels of phytosociological classification
bór brusznicowy – cowberry coniferous forest
bór czernicowy – bilberry coniferous forest
bór bażynowy – crowberry coniferous forest
ols porzeczkowy – red currant swamp alder forest

Things become more complicated when otherwise convenient names, such as *grąd* or *łęg,* are subdivided. Okołów and Okołów (2005), following the phytosociological classification, split the alliance of *grąd*, association of *grąd subkontynentalny*, in the Białowieża Forest into five subassociations: *grąd czyśćcowy, grąd typowy, grąd miodownikowy, grąd turzycowy* and *grąd trzcinnikowy*. Most of these are derived from the name of the plant, the presence of which enables the fine distinctions to be made. The English names of these plants can be checked and incorporated in the name of the entire habitat. Since English imposes the necessity of knowing more basic characteristics concerning tree composition to attach an appropriate label to a habitat, one can fall into a trap. It turns out that it is not enough to say hedge woundwort oak-hornbeam forest because the chief distinction is that it grows in lower, wetter situations and has an admixture of ash trees. Besides, in the Białowieża Forest the lime constitutes a vital ingredient of *grąd*. To be able to properly differentiate between the types the English names reach impressive lengths. Thus, *grąd czyśćcowy* should be hedge woundwort oak-hornbeam-lime-ash forest and its driest variety *grąd trzcinnikowy* – small reed oak-hornbeam-lime-spruce forest. Incidentally, if the fact that 'lime' is often called 'linden' is taken into consideration, the alternative name emerges in each case, hostile as it is to the idea of eliminating ambiguity and multiplicity in terminology.

Various guidebooks and brochures tend to contain a lot of informal or original habitat divisions that have never been officially translated into English so as to set a standard. What is more, Polish official documents introducing new standards take advantage of these divisions in order to establish a point of reference for the sake of clarity. For example, *Decyzja nr 5 30.01.2007,* the document issued mainly in connection with the Natura 2000 programme, instructs foresters on the recognition of *grąd połęgowy* – a habitat which is not distinguished in *Poradniki ochrony siedlisk i gatunków,* the work approved by the government and scientific circles. Literally, *grąd połęgowy* is oak-hornbeam forest on no longer flooded riparian forest sites. With some probability, it may be the G1.A1 EUNIS classification habitat. Were it to be confirmed, the English equivalent would be: *Quercus–Fraxinus–Carpinus betulus* woodland on eutrophic and mesotrophic soils. Both names, informative as they are, look rather awkward. This may well be the case of culture-specific items that in one language are well-absorbed and take short verbal forms, and in another are alien and can only be described with the help of several words. This being an intriguing question, the establishment of links between culture and nomenclature as well as terminology concerning nature is attempted in Chapter Three. To quote further examples, this time from *Poradniki ochrony siedlisk i gatunków*, translators might have to grapple with *grąd niski* and *grąd wysoki*, a source of many a mistranslation (see

Chapter Four). The importance of standardisation of expressions such as *grąd niski* or *grąd wysoki* for the correct understanding of correlations between Polish and international nomenclature within the Natura 2000 programme is almost none. However, such a standardisation becomes highly useful for translators of popular publications on nature, where informal terms and simplified or informal classifications abound.

2.3. The intra- and interlingual translation of non-forest habitats

The intralingual Polish–Polish translation of non-forest natural habitats for Natura 2000 purposes begins with a different challenge than in the case of woodland types. As demonstrated by Pawlaczyk and Mróz (2003), forestry documents either do not record them, treating them as part of a woodland habitat, or conceal them under vague, unscientific terms such as *bagno* – marsh, bog, swamp, fen, morass, quagmire, mire, *nieużytki* – wastes, wasteland, uncultivable land, *piaski* – sands, *halizna* – fallplace, blank (Czekierda 1992), an open treeless area created by a natural disaster or processes such as hurricanes or raised water table, *płazowina* – irregularly stocked open stand, *pastwisko* – pasture *or łąka* – meadow. *Wrzosowiska* (heaths), in addition to being marked as *bagno, nieużytki, halizna* or *płazowina*, sometimes function under their own name, but they may just as well signify stands of small reed (*Calamagrostis epigeios*). Like various kinds of *murawy* (xerothermic grasslands), they may occur in patches at the edge of tree stands or in gaps in them and as such do not exist, at least officially. *Łąki* usually cover all sorts of meadow habitats, both protected within the Natura 2000 programme and not protected, but may hide alkaline fens (habitat 7230). *Pastwisko* can simply be a grazing pasture but may stand for e.g. *murawy bliźniczkowe* (habitat 6230 species-rich *Nardus* grasslands, on siliceous substrates in mountain areas (and submountain areas in Continental Europe)). *Piaski*, literally sands, can be actually identified as habitats: 2330 inland dunes with open *Corynephorus* and *Agrostis* grasslands, 6120 Xeric sand calcareous grasslands or initial stages of 91T0 Central European lichen pine forests.

The most impressive collection of valuable habitats is labelled as *bagno*: 7110 active raised bogs, 7120 degraded raised bogs still capable of natural regeneration, 7140 transition mires and quaking bogs, 7150 depressions on peat substrates of the *Rhynchosporion*, 3160 natural dystrophic lakes and ponds, 7230 alkaline fens, 7210 Calcareous fens with *Cladium mariscus* and species of the *Caricion davallianae*, 7220 petrifying springs with tufa formation (*Cratoneurion*), 6510 lowland hay meadows (*Alopecurus pratensis*, *Sanguisorba officinalis*) and 91D0 bog woodland. But some habitats not included in the Habitats Directive, such as reed or tall sedge swamps, may be found in *bagno* as well.

Generally speaking, in order to translate the State Forests documentation of non-forest habitats into Natura 2000 conservationist language, the naturalist must allow for a wider variety of possible equivalents per item. While in the case of woodland types the question is narrower (Which type of forest?), here it is rather broad (What kind of landscape?). A site registered as woodland does not usually turn out to be an open landscape, but an area expected to instantiate just that, marked *piaski* or *bagno*, might appear to support a young pine growth without a stretch of sand dunes or marsh. The linguistic view of the world emerging from forestry documentation departs drastically from the reality. Specialist linguistic categories are as fuzzy as folk ones.

Moreover, the Polish names of Natura 2000 non-woodland habitats, especially mires, have changed remarkably over the five years since they received their first official translation in Dyduch-Falniowska (2002). One of the reasons for this may be the lack of straightforward correspondence between the names used for Natura 2000, recommended by international mire conservation bodies and functioning in Polish and British scientific circles. The most simplified categories of mires in Polish are *torfowisko wysokie*, *torfowisko przejściowe* and *torfowisko niskie*. The English equivalents are: raised bog, transition/al mire and fen respectively. Their Polish names derive from the way water is fed and from the locations they are found in – *wysokie* (literally: high) get solely nutrient-poor water from precipitation and occupy basins between elevations, *niskie* (low) take advantage of underground mineral-enriched water and lie in lower places within extended river valleys, and *przejściowe* combine the features of the former two. Any other significant characteristics, such as whether they are acidic or alkaline, fed by flowing or spring waters, etc. are ignored in this division. The English names resemble the Polish ones except that 'fen' is borrowed from general English, in which it denotes low-lying, nutrient-rich wetland. A more comprehensive ecological typology of peatland, given by Ilnicki (2002: 186), consists of five major types:

1) *torfowiska wysokie* (raised bogs)
2) *torfowiska przejściowe* (transition mires)
3) *słabo kwaśne torfowiska niskie* (low acidity fens)
4) *alkaliczne, bogate w wapń torfowiska niskie* (alkaline, calcium-rich fens)
5) *eutroficzne torfowiska niskie* (eutrophic fens),

and in terms of vegetation cover, these correspond to:

1) *mszary ombrotroficzne* (ombrotrophic sphagnum moss bogs)
2) *mszary minerotroficzne + silnie kwaśne mechowiska* (minerotrophic sphagnum moss bogs + strongly acidic sedge-moss fens)
3) *kwaśne młaki i mechowiska* (acid flush fens and sedge-moss fens)
4) *mechowiska kalcyfilne* (calciphilous sedge-moss fens)
5) *szuwary turzycowe + szuwary właściwe + olsy* (tall sedge swamps + reed swamps +alder swamp forests).

Two of the basic categories, identical with the first two Ilnicki types, appear in the Polish version of the Natura 2000 habitat list. *Torfowisko niskie*, which Ilnicki pays so much attention to, is absent and replaced by a long label containing names coined on the basis of hydro-geological and phytosociological qualities. This is probably due to interpretive problems concerning the very definition of code 7230.

Pawlaczyk and Mróz (2003: 75) warn that the entry *bagno* can conceal habitats both protected within the Natura 2000 programme and not protected therein. This statement is followed by the specification of those left outside it: *turzycowiska, torfowiska niskie, zdegradowane torfowiska i inne mokradła.* Several years have elapsed and *turzycowiska* have entered the Natura 2000 exclusive club. At least they appear in *Poradniki ochrony siedlisk i gatunków* in the Polish name of code 7130: *górskie i nizinne torfowiska zasadowe o charakterze młak, turzycowisk i mechowisk* – montane and lowland alkaline fens in the form of flush vegetation, sedge beds and moss beds. Normally, the name *turzycowisko*, which derives from *turzyca* (sedge) refers to tall sedge swamp, flooded by non-alkaline river waters – the non-Natura 2000 habitat. But it can also denote fragments of fenland interspersed with tall sedges and small sedges, small sedge-moss assemblages being labelled *mechowisko*. What must have happened is that Pawlaczyk and Mróz (2003) used the former meaning, while the other source, the latter. Needless to say, it contributes to terminological chaos and a sense of alienation among interested non-specialists. This sense intensifies with a discovery that *torfowisko niskie* overlaps with alkaline fen and, to some extent, with code 7140, gathering transition mires and quaking bogs. Therefore, whatever its subtype, it is always guaranteed protection within the Natura 2000 network. The only habitat often interpreted as fen, which stays outside the Natura 2000 programme, is *turzycowisko*, but Pawlaczyk and Mróz (2003) put it next to, and not within the *torfowisko niskie* category. Even more surprisingly, *zdegradowane torfowiska* (degraded peatlands), which these authors refuse to grant a Natura 2000 conservation status, have it, either within 7120 or as one of the meadow habitats, especially 6410 and 6510. Unless turned into turbaries or destroyed beyond regenerative capacities by total drainage and the application of lime, degraded peatlands are protected. The question arises whether all those discrepancies are a matter of lack of precision, clumsy formulation, a misprint or a reflection of varying interpretations, gradually modified and compromised.

Irrespective of these speculations it must be noticed that in a number of cases the impact of negotiations over the meaning seems unquestionable. Wherever an English name's capacity for interpretation is low, Polish labels have remained identical throughout, like, for instance, *torfowiska wysokie* (raised bogs). Where it is higher, as in the case of 7140 transition mires and quaking bogs, 7150 depressions on peat substrates of the *Rhynchosporion*, and above all, 7230 alkaline fens, labels have changed. Below follows a short analysis of these reformulations. For the sake of convenience the four sources are referred to as **A** – Dyduch-Falniowska et al. (2002)**, B** – Pawlaczyk and Mróz (2003)**, C** – *Poradniki ochrony siedlisk i gatunków* and **D** – http://www.lkp.org.pl/poradniki/lasy/fotoprzewodnik_nielesne.pdf**. A** and **B** agree in the three cases: 7140 is *torfowiska przejściowe i trzęsawiska*, 7150 *obniżenia dolinkowe i pła mszarne*, and 7230 *torfowiska alkaliczne*. Some time has passed and **D,** having taken away *pła mszarne* from 7150, calls 7140 *torfowiska przejściowe, pła i trzęsawiska*, and labels 7150 as *przygiełkowiska,* which means 'beak sedge beds'. 7230, in turn, is *torfowiska alkaliczne (mechowiska)*. The word in brackets indicates an evolution towards specification of what can be included in a very broad category of alkaline fens. In **C** 7230 precision reaches a high level indeed: *górskie i nizinne torfowiska zasadowe o charakterze młak, turzycowisk i mechowisk* – montane and lowland alkaline fens in the form of flush vegetation, sedge beds and moss beds. 7150 is *obniżenia na podłożu torfowym ze związku Rhynchosporion*, which is a faithful word-for-word translation of the English version and 7140 returns to its previous shape: *torfowiska przejściowe i trzęsawiska.*

Since mires can be differentiated along a range of gradients (fertility, water table, acidity- alkalinity, etc), it is not easy to accommodate the resulting wealth of types within the categories imposed by the Habitats Directive. Moreover, researchers disagree as to the usefulness of certain gradients with the effect that not all the labels have currency everywhere in Europe. An overview of European schools of thought on the subject will be presented later. First, some space will be devoted to the Polish scene. The main problem is that Polish categories do not fit neatly into the Natura 2000 code names. *Torfowiska niskie otwarte* (open fens) belong both to 7230 and 7210. 7140 embraces mainly 7210, 7230 *mechowiska* (sedge-'brown moss' fens) and *mszary* (sphagnum moss bogs), especially 7150. *Mechowiska* require significantly different acidity conditions from *mszary* and support distinct vegetation: 'brown moss'-dominated versus sphagnum moss-dominated, and yet they fall into the same category. Thus, 7140 covers a borderland between fens and transition mires, since quaking mats of vegetation can develop around oligotrophic, acidic lakes as well as over alkaline water-fed loose unconsolidated peat. The *Interpretation manual* itself, where 7140 is 'transition mires and quaking bogs', explains that 'small fens between water (lakes, ponds) and mineral soil' are included. Bog and fen put together is somewhat contradictory. The Polish version at least avoids such nomenclatural contradictions. The broad terms *torfowiska przejściowe i trzęsawiska* can accommodate both bogs and fens without an implication of illogicality or internal incoherence, however unwelcome this might be from an ecological point of view. Furthermore, 7150 groups raised bogs and transition mires, with emphasis placed on hollows in the former and floating mats in the latter. The hollow-hummock structure of raised bogs as a whole is protected under the 7110 code. This division looks odd, but supposedly it aims to focus attention on the appropriate conservation of patches of sphagnum moss, preferably with beak sedges, among fens. 7110 and 91D0 represent another category with blurred boundaries. The *Ledo-Sphagnetum Magellanici* association can be interpreted both as raised bog and bog woodland, overgrown with sparse stunted pines. Incidentally, bog woodland may be a further source of misunderstandings because, as just pointed out, it can support plant communities adapted to nutrient-poor, acidic conditions (bogs and bog woodland) as well as fen carr, such as *las brzozowo-sosnowy – biel,* developing in nutrient-rich, alkaline fens.

The already quoted Internet guide[27] summarises correspondences between mire types widely used in Poland and Natura 2000 categories as follows:

Torfowisko wysokie – 7110, 91D0
Trzęsawisko – 3160, 7140, 91E0, 7210
Torfowisko przejściowe – 7140, 91D0
Torfowisko źródliskowe – 7220, 7230, 91E0

[27] http://www.lkp.org.pl/poradniki/lasy/fotoprzewodnik_nielesne.pdf

Torfowisko mechowiskowe – 7230, 91E0
Torfowisko zalewane – 91E0

As can be seen, some mires can be protected only when already transformed from an open into a woodland habitat. *Torfowisko zalewane* (floodwater mire), when in the form of reed swamp or tall sedge swamp, cannot count on a Natura 2000 status, but as an alluvial forest it can. Some others can be easily mistaken for a non-Natura 2000 habitat: *torfowisko mechowiskowe* may look like tall sedge swamp or even as reed swamp, but mosses at the base of reed stems prove it is *mechowisko*. Similarly, *trzęsawiska kłociowe* – quaking mats of saw grass in calcareous fens (7210) may be surrounded by reeds and thus ovelooked by people performing field surveys. But even if wrong identification does not lead to denying priority conservation status, the consequences can be disastrous. If, for example, a sedge-moss fen (*mechowisko)* is recorded as '6510 lowland hay meadow', because somebody has failed to check the wettest parts of a site, with the 'brown moss' layer, the conservation recommendations may be as for a meadow. In the case of hay meadows European Union agri-environmental schemes, which aim to promote active conservation by extensive management, allow the application of fertilisers. A farmer is then paid to maintain a less valuable habitat at the cost of the one whose role in enriching biodiversity is more important and cannot be played in excessively nutrient-rich conditions. The payments for sedge-moss fen management are available in exchange for the obligation not to use fertilisers and to mow without heavy equipment. Of course, drainage is out of the question, too.

Remarkably, the Polish categories quoted at the beginning of the above paragraph stem from a variety of aspects. In *torfowisko wysokie* and *przejściowe* what happens to be enhanced is hydrological and topographical factors, in *trzęsawisko* the peat formation strategy, in *torfowisko źródliskowe* and *torfowisko zalewane* the origin of the water, and in *torfowisko mechowiskowe* the dominant vegetation. The list of possible names is much longer, depending on the angle from which mires are studied. Phytosociology has contributed several names which bring the plant cover to the fore: *mechowisko*, *mszar* (e.g. *kępkowy, dolinkowy, przygiełkowy*), *młaka* (*niskoturzycowa*) and *turzycowisko* (e.g. *mszyste, trawiaste*). Hydro-geological considerations have given rise to names such as *źródlisko, torfowisko przepływowe, torfowisko nakredowe* or *torfowisko alkaliczne, młaka* – which is flush vegetation but, at the same time, also a flush itself (a wet area associated with freshwater springs).

The subtypes of Natura 2000 categories distinguished in *Poradniki ochrony siedlisk i gatunków* follow the same path of pragmatic eclecticism. They do not keep to one selection criterion. Neither do they reflect the categories which figure in the PHYSIS-CORINE database. Within 7230 alkaline fens three smaller units are differentiated: 1) *młaki górskie* (montane flush fens), 2) *torfowiska zasadowe Polski południowej (z wyłączeniem gór) i środkowej* (alkaline fens of southern (without mountains) and central Poland) and 3) *torfowiska źródliskowe i przepływowe Polski północnej* (spring and percolation fens of northern Poland). Apart from the criteria already mentioned the altitude and geographical location come into view here. The PHYSIS code placed under 7230 is 54.2 (rich fens), which implies that it encompasses twenty-nine subtypes – nearly ten times more – such as black bog-rush fens, Davall sedge fens, black sedge alkaline fens, small herb alkaline fens. Three subtypes of 7110 also underline the geographical distribution: *niżowe* (lowland), *sudeckie* (Sudety mountains), *karpackie* (Karpaty mountains) *torfowiska wysokie* (raised bogs). Again, no direct transfer of PHYSIS categories has been performed since code 51.1 comprises thirty-one subtypes, for instance, bog hummocks, ridges and lawns, bog hollows, bog pools, bog seeps and soaks, lagg, bog pre-woods.

In *Poradniki ochrony siedlisk i gatunków* all the other mire habitats have just one subtype each. These are identified through a set of diverse criteria. In three cases this number of subtypes coincides with the number of subtypes within the corresponding PHYSIS code: 7120 – 51.2 purple moorgrass bogs, 7150 – 54.61 white beak sedge communities, 7220 – 54.12 hard water springs. But 7210 corresponds with the two PHYSIS subtypes: 53.31 fen Cladonium beds and 54.2 (rich fens), while 7140 with code 54.5, which groups twenty-four subtypes such as slender-sedge swards, sphagnum and cottongrass rafts, brown moss carpets, and Alpine deergrass quaking bogs. The subtypes of the PHYSIS-CORINE classification do not have any official translations into Polish. In Dyduch-Falniowska and Zając (eds) (1996), Polish translations or descriptive interpretations exist for barely a handful of biotopes. Subtypes may not even have the Latin name. The correct rendering of, for example, bog seeps and soaks requires an examination of the particular site and consultation with a phytosociologist. It appears that bog seeps and soaks can become *niższe kępki i płytsze dolinki* provided that the context rules out the possibility of the misinterpretation of such ambiguous words as *kępki* and *dolinki*.

Parallel to Polish, a large quantity of English names of mire habitats circulate: rich fen, poor, calcareous fen, sedge-moss fen, quaking bog, sedge beds/swards, springs, hummocks, hollows variously qualified by characteristic plant names, etc. The most impressive collection can be found in the EUNIS list of habitats, which endeavours to register all landscape types of the Palearctic – 324 mire communities. In fact, the number of labels may be greater than in Polish since a lot of authors from countries other than the UK write about peatlands in English, not to mention numerous international documents drafted in English. In Europe there are two extremely active schools of thought: north-western with its Scandinavian variety and Central European, again not exactly uniform. North-western classifications describe mires in terms of bog, fen and swamp. Transition mire is just a translation of a category distinguished by Central European researchers and employed in the Habitats Directive. Westerners differentiate mire types on the basis of two factors: 1) alkalinity, inextricably bound with calcium content, pH and hydraulic conductivity (level of minerals) and 2) fertility – phosphorus and nitrogen content, Central Europeans on the basis of dominant plant species. The drawback of the latter is that plants mix in different ways and in varying proportions, forming multiple alliances and associations. If treated as separate types, their number would be overwhelming. This situation precludes making valid generalisations concerning other important properties of mires. For instance, the long-lived assumption that the presence of sphagnum mosses indicates acid conditions, has proved untrue after the examination of a sufficiently wide range of combinations. That is one of the reasons why it is so difficult to devise a classification, in which one category would relate to one concrete mire site and could thus provide exhaustive and unambiguous information on its conservation needs. Classifications based on more thorough investigations help take more conscious and cautious conservation decisions but they by no means eliminate dilemmas such as whether to preserve a mire as a whole or its particular fragments, or whether to protect the actual or potential plant cover.

The different approaches necessitate translation from system to system. The following correspondences can be drawn: transition mire can be classified as poor fen or minerotrophic bog, rich fens are those calcium-rich as well as those base-rich (e.g. including magnesium and sodium) – in the language of Natura 2000 rich fens encompass alkaline and calcareous fens. Bogs are bogs – ombrotrophic and acidic. Wheeler and Proctor (2000: 199) presented a relationship between these basic types of bog, fen and swamp and vegetation types, and, generally, they attempted to minimise the quantity of existing labels. They did it as an offshoot of seeking consensus on the validity of certain ecological gradients in north-west European mires. A noble idea in itself, it soon lost much of its appeal when Økland, Økland, and Rydgren (2001) responded that some of the gradients played down by Wheeler and Proctor (2000) were regionally very important and as such could not be ignored. No matter the details, the effect is that, for example, they insisted on extra categories of fen: 1) along the gradient from acid, base-poor to base-rich: poor fen fertility (subdivided into extremely and moderately poor), intermediate fen and rich fen (again extremely and moderately) 2) along the gradient from spring to flush to fen: sloping fen and level fen. Some of those proposals dwell on the studies by Sjörs and his successors. Sjörs (1950)'s division along the base-richness gradient started from moss (the older term for bog) and went through extreme poor fen, transitional poor fen, intermediate fen, transitional rich and rich fen. If one adds the Finnish contribution (Ruuhijarvi 1983), with names such as concentric and eccentric bogs, plateau bogs, aapa fens, palsa mires, and a long list of habitats based on vegetation units, such as hummock and hollow pine bog or herb-rich palludified spruce forest, or British/Irish blanket bogs, or types determined by topography: valley mires, valleyside mires saddle mires, spur mires, basin mires or watershed mires (Parkyn, Stoneman, and Ingram (eds) 1997), condensation mires, kettle mires (Joosten 2001), a thicket of categories becomes dense indeed. Obviously, their boundaries criss-cross one another and in each category there are better and worse exemplars.

Although scientists have been trying to achieve more uniformity in nomenclature, the diverse needs of conservation, also with regard to regional variation, cause the multitude of names to stay put, or even increase. Inevitability of such diversity is confirmed by Joosten (2001, unpaginated): 'a detailed typology should be based on a combination of various criteria … e.g. hydro-genetics, trophy and base-saturation conditions, macro-morphology, micro-morphology, floristics, vegetation, physiognomy, and stratigraphy'. An example of a combination can be 'percolation mires found in mesotrophic acid, mesotrophic subneutral, and mesotrophic calcareous variants' (ibidem). Joosten and Clarke differentiated mire types on the basis of several criteria and concisely arranged them in a table (2002: 40), later somewhat modified (Couwenberg 2005). Some names were coined, the majority adopted and further subdivided. Therefore some can be elegantly rendered in Polish, for example, blanket bog – *torfowisko kołdrowe* or *torfowisko wierzchowinowe* (a Polish variety of very limited distribution), percolation mires – *torfowiska przepływowe*, floodwater mire – *szuwary nadrzeczne*. Many others do not have official translations and, if translated, they will sound as technical as they do in English: ombrogenous surface flow mire (*torfowiska ombrogeniczne o powierzchniowym przepływie wód*?), lithogenous immersion mire (*litogeniczne torfowiska pływające* or *pła litogeniczne*), thalassogenous acrotelm mire (*talasogeniczne torfowiska wysokie*?), etc. Probably as a way to avoid sounding hermetic some applied science works prefer more descriptive, though less elegant, terms such as groundwater fed mires (*torfowiska zasilane wodą gruntową*) instead of soligenous mires (*torfowiska soligeniczne*) (e.g. Grootjans and Wołejko 2007: 22).

Putting terminology in order is a messy and long-lasting process since even newly published dictionaries do not keep up with the changes introduced by scientists. Besides, their old versions may still be used long after successive updates have appeared. Table 1 illustrates incongruities between up-to-date terminology and entries for the term *torfowisko* in five bilingual dictionaries.

Table 1: Incongruities between up-to-date terminology and entries for the term *torfowisko* in five dictionaries.

Polish term	Ilnicki 2002	Current internationally agreed English term	Bick et al. 1976	Czekierda 1992	Metera and Pezold (eds) 2003	Stanisławski 1989	Linde-Usieknkiewicz 2004
Torfowisko	Peatland/mire	Peatland/mire	In one entry with *bagno*: bog, fen, mire, moor, peatland	Peatbog, peat swamp, peat-moor, moor, peatland,	Peat bog	Peatbog, turbary	Peat bog
Torfowisko niskie	Fen, eutrophic mire	Fen	Fen, low moor, low bog	Lowmoor, lowland bog	Low, peat bog, fen	-	Lowland bog
Torfowisko wysokie	Raised bog, bog, oligotrophic mire	Raised bog	High moor, raised bog, raised moss	Highmoor, raised bog, high bog, highmoor, peatbog	High peat bog	-	High peat bog
Torfowisko przejściowe	Transitional bog, mesotrophic mire	Transition mire	Transition bog/moor	Transitional moor	Transitional peat bog	-	-

Although Ilnicki (2002) offers internationally agreed current terms, he does not resist temptation to include one or two alternative interpretations. Since there are more interpretative varieties available (e.g. *torfowisko przejściowe* can be minerotrophic bog or poor fen), his choice seems arbitrary. Therefore the third column is added to pinpoint the most exact internationally accepted equivalent of the Polish term. 'Turbary' refers to a peat-extraction site and is hardly ever used. As a denotation of the living system the dictionaries give many possibilities. Czekierda (1992) and Bick et al. (1976) provide a long list of equivalents, most of them obsolete (moor, raised moss) or incorrect. Despite the fact that 'peatbog' implies 'raised bog', the type essentially different from 'fen', it is indicated in most dictionaries as the umbrella term for all the types. Only 'peatland' and 'mire' are currently accepted as broadly applicable, the first being the most comprehensive: 'an area with or without vegetation with a naturally accumulated peat layer at the surface', the second: 'a peatland where peat is currently being formed and accumulating' (Joosten and Clarke 2002, see also Bragg and Lindsay (eds) 2003). 'Fen', already included in Bick et al. (1976), is ignored in most dictionaries, although this is the scientifically agreed and widely used term meaning *torfowisko niskie*. The presentation of 'lowland bog' as 'fen' deserves a particularly strong dose of criticism because it signifies **raised bog** located in the lowlands as opposed to those found in the mountains. Regrettably, newer publications tend to glean terms from old specialist dictionaries, reflecting the abundance of uncoordinated scientific research but overlooking later endeavours to standardise terminology. Thus, the translator must be up-to-date with specialist literature and cannot trust even the seminal books of the past such as *The peatbogs of Polesie* by Stanisław Kulczyński (1949), which misleadingly suggests to the modern recipients that Polesie was or is covered solely by raised bogs and no fens.

Many authors admit that peatland terminology is in a state of confusion (Bragg and Lindsay (eds) 2003, Hofstetter 2000, Ilnicki 2002, Joosten and Clarke 2002, Wheeler and Proctor 2000) and international conferences attended by mire specialists have been held in order to formulate agreed definitions and unequivocal labels. Hofstetter's proposal, modelled on the international FAO-UNESCO soil classification, goes as far as to introduce artificial terms. They derive from Greek and Latin stems and suffixes, and as such can only be employed for scientific purposes. Translation for the sake of applied studies and widespread education will be indispensable. As in the case of soil types some official documents would use old nomenclature for a long time, some others would gladly embrace the novelty. In short, chaos stands little chance of elimination, although communication within a narrow group of specialists may improve. To sample Hofstetter's idea it suffices to quote a few newly-coined terms:

> Terruds (*terrus* = earth, *udus* = damp, moist, wet) –
>
> communities and ecosystems in which the presence of water in the uppermost edaphic/pedic layers is of sufficient duration to promote a) anaerobic conditions long enough during the growing season to effect perennial selection of the dominant plants or animals and (or, in the the absence significant biota) b) persistent anaerobic edaphic indicators, but not consistently so deep as to exclude the establishment of potential, perennial, emergent macrophytes, and that are transitional between terrestrial (upland) and aquatic communities and ecosystems. (Hofstetter 2000, unpaginated)

The term is meant to cover wetlands but understood less broadly than the Ramsar definition[28], which encompasses even natural and artificial freshwater and saltwater bodies. Various features of terruds are determined with any number of adjectives. For example, one can end up with 'atransfluo hypotrophic aeriohydric sphagnophytic insito-sphagnopedic terrud', which stands for 'an ecosystem transitional between terrestrial and aquatic, lacking or almost lacking horizontal water flow, containing only small amounts of nutrients, fed by water from the atmosphere, characterised by sphagnum mosses, which simultaneously serve as the main ingredient of peat produced on the spot.' This corresponds to 'raised bog'. As such it may seem an exaggeration but it must be remembered that terms applied across the globe do not always closely overlap, and an exact equivalent of raised bog may not exist in some languages. Furthermore, in all sorts of transitional cases Hofstetter's tools facilitate unequivocality and effective international communication. On the other hand, the proponent realised that new terminology was cumbersome and entirely different from what everyone had been accustomed to, so it would be potentially slow to disseminate and gain worldwide currency. Predictably, since the year 2000 it has not stimulated much practical support in international circles.

Contrary to common belief and the wishes of terminologists the signs of specialist languages are not at all precise and unequivocal. It is the fact that specialist languages describe a fragment of reality presented as a model that makes them so (Hullen in: Skowronek 2001: 117). Since,

[28] The text of the Ramsar Convention on Wetlands, signed in 1971, is available at http://www.ramsar.org

however, the models are numerous and use the same signs to denote different items, the signs often require disambiguation. In other words, the meaning of the same term or name varies with a scientific discipline. For instance, in botany *krwawnik* is a white-flowered perennial (yarrow) but in geology it is a blood-red mineral cornelian; similarly in zoology *kukułka* is a bird (the cuckoo), in botany it is one of a number of *Dactylorhiza* orchids; *łęg* in phytosociology and forestry is a kind of nutrient-rich broadleaved forest formed in areas repeatedly flooded by rivers or kept wet by flowing underground waters, in grassland management it is a type of regularly inundated meadowland (for the last one I follow Wołejko et al. 2004). Alarmingly, within one discipline authors endow the same term with different meanings. 'Soligenous' illustrates the point well. According to Joosten and Clarke (2002: 32) it used to be defined as originated not only by precipitation but also 'by meteoric water running off from the surrounding terrain', then as 'originated under the influence of streaming groundwater' and, simultaneously, as 'originated solely under the influence of spring water'. In mire studies the origin of water matters a lot and the existence of such discrepancies may baffle. Lack of uniformity in the definition of absolutely basic terms such as peatland or mire does not just surprise, it shocks. While Bragg and Lindsay (eds) (2003), Joosten and Clarke (2002), and Parkyn, Stoneman, and Ingram (eds) (1997) agree that mires are peatlands where peat is currently being formed and accumulating, and that peatlands where peat accumulation has stopped are no longer mires, Wheeler and Proctor (2000), following Mornsjö, suggest that 'mire' should mean peat- and some mineral-based (tufa, humus) wetlands, which corresponds to Hofstetter's 'terrud'.

In other words, Bragg and Lindsay (eds) (2003), Joosten and Clarke (2002), and Parkyn, Stoneman, and Ingram (eds) (1997) regard mire as a subcategory of peatland, whereas Wheeler and Proctor (2000) make peatland a subcategory of mire. In both interpretations the membership of the category of peatland remains the same. Thus, mire, incorporating some of the members of the category of wetlands, becomes a far broader category than peatland has been in either case. The above is a good illustration of the stretchability of categories and their dependence on background framing. It also demonstrates how the nature of categorisation impedes the process of intralingual translation of nature terminology.

2.4. The intra- and interlingual translation of categories at the level of species

So far I have dealt with challenges nature conservationists and scientists meet, trying to differentiate and classify larger units – habitats or plant alliances and associations. Few doubt this is a difficult task. With so many factors at play, countless combinations of species dependent on various conditions such as water and nutrient availability, structure of soil, etc. any organisational undertaking simply must entail tremendous effort. As to classification at the level of species, popular opinion veers towards assuming that everything has been neatly arranged by Carl Linnaeus, with Latin serving as the lingua franca and solving all the problems of communication between speakers of different modern languages. Not many people realise that Latin names have synonyms, not all of which are always given in all sources. Some of them seem to be less favoured than others or treated as defunct in most publications but occasionally revived in one or two. Most importantly, new discoveries have necessitated constant changes in taxonomy and the rapid progress of genetics keeps revolutionising the views of systematicians. Hence one species becomes split into two, they may shift into different families, and all the other taxa (order, genus, phylum, etc.) are regrouped, completed, renamed, degraded or upgraded in status. National nomenclatures also change, not necessarily just to reflect those revolutions in the Latin system. The number of names in circulation increases. Names swap referents, and sometimes shortened, become more obscure.

Synonymy in Latin nomenclature comes from the fact that in the past scores of species were described and named independently by various authors. Information spread slowly and, as a result, names of species far surpass their actual number (Szujecki 1998: 29). Here are some examples from Anioł-Kwiatkowska (2003):

- *goździcznik wycięty* (proliferous pink) – *Tunica prolifera, Petrorhagia prolifera, Dianthus prolifer, Kohlrauchsia prolifera,*
- *czyścica storzyszek* (wild basil) – *Clinopodium vulgare, Calamintha clinopodium, Calamintha vulgare, C. vulgaris, Satureja vulgaris,*
- *włosienicznik tarczowaty* (common water crowfoot) – *Batrachium peltatum, B. dichotomum, B. langei, B. trichophyllos, B. peltiveri, Ranunculus aquatilis,*
- *podejźrzon księżycowy* (moonwort) – *Botrychium lunaria, Osmunda lunaria.*

Naturally, authors seldom put all the names together. Therefore in vain would one look for *Tunica prolifera* in Blamey, Fitter, and Fitter (2003). The proliferous pink is there solely *Petrorhagia prolifera.* Similarly, the wild basil is presented as just *Clinopodium vulgare* and so forth in the same vein. Many a search ends fruitlessly due to the shortage of time, which adversely affects the quality of communication, worse still if translation is involved.

Double nomenclature occurs also at the order level: *Orthoptera = Saltatoria, Neuroptera = Planipennia, Siphonaptera = Aphaniptera* (insects). International regulations give priority to the first useful name ever registered, while others are allocated the role of synonyms. Modern language nomenclatures do not adhere to this principle. Uniformity and consistency can only be appealed for (Szujecki 1998: 30). Partially because of that, synonymy in modern languages does not necessarily reflect the one in Latin: although like in Latin, *Orthoptera* have two equal names in Polish: *prostoskrzydłe = szarańczaki* and a very long name in English: grasshoppers, locusts, and relatives, *Neuroptera = Planipennia* are just *siatkoskrzydłe* (though in some publications they are *sieciarki).* However, *Thysanoptera* (thrips) are *przylżeńce = wciornastki* and *Psocoptera* (booklice and barklice) either *gryzki* or *psotniki.* In the plant kingdom the family *Cyperaceae* (sedge family) can be *ciborowate = turzycowate*, *Boraginaceae* (borage or forget-me-not family) *szorstkolistne = ogóreczníkowate.* The subclass *Bryidae* (no English name) has earned three names *mechowcowe = prątnikowcowe= mchy właściwe* (Jasnowska, Jasnowski, and Radomski 1995), informally but consistently and frequently called *mchy brunatne* (brown mosses) as opposed to *torfowce* – sphagnum (bog) mosses[29].

It sometimes happens that a Latin name used in a publication seems to be absent from other sources, already proven reliable and detailed, and detective work must be done to figure out the actual referent. *Salix atrocinerea* is a valid example. Mentioned by Evans (1992) as identical with 'the sallow', it is virtually absent from Anioł-Kwiatkowska (2003) and Blamey, Fitter, and Fitter (2003). Consulted botanists have also been clueless. Fortunately, in the USDA PLANTS database[30], it has turned out to be the large gray willow. It has the following Latin synonyms: *Salix cinerea ssp. atrocinerea, Salix cinerea ssp. oleifolia, Salix oleifera.* In Anioł-Kwiatkowska (2003) *Salix cinerea* (*wierzba szara*) is 'gray/grey

[29] See e.g. *Poradniki ochrony siedlisk i gatunków* at http://natura2000.mos.gov.pl/natura2000/pl/poradnik.php - 86k or http://www.lkp.org.pl/poradniki/lasy/fotoprzewodnik_nielesne.pdf).

[30] Available at http://plants.usda.gov/java/profile?symbol=SAAT2 - 43k

willow' or 'common sallow'. Before one can breathe a sigh of relief, though, a glimpse at the relevant entry in Blamey, Fitter, and Fitter (2003) ruins the sense of satisfaction: 'common sallow' is *Salix caprea* = goat willow (*wierzba iwa*), *Salix cinerea* being 'grey willow/sallow'. In sum, according to two specialist sources 'common sallow' can signify two separate species. The inventory of complications does not end there: what one book considers synonyms, elsewhere may appear as names of different species. In Anioł-Kwiatkowska (2003) *turzyca delikatna* is *Carex disperma, Carex supina,* and *Carex tenella.* Kaźmierczakowa and Zarzycki (eds) (2001) have separate entries for *Carex disperma* and *Carex supina*, *Carex tenella* being a synonym of *Carex disperma*, whose Polish name is *turzyca szczupła*. Only *Carex supina* is called *turzyca delikatna.* Blamey, Fitter, and Fitter (2003) do not describe these two sedges, which indicates that they do not occur in Britain and Ireland. American sources come to the rescue. We learn from the USDA PLANTS database[31] that, depending on the state, *Carex disperma* (*turzyca szczupła*) is 'softleaf sedge, soft-leaved sedge, two-seeded sedge or two-seeded bog sedge'. *Carex supina* (*turzyca delikatna*) is 'weak arctic sedge'. Sadly, the name 'soft sedge', which Anioł-Kwiatkowska (2003) gives to both species, seems not to belong anywhere.

Undeniably, in spite of synonymy, Latin often does help, if provided. Guidebooks and brochures tend to keep to native names, chosen at will rather than according to a criterion of official recognition. To exemplify, Okołów and Okołów (2001: 42) write about *majownik dwulistny*, a name either regional or obsolete, obscure to the scientists specialising in the field. With no Latin names given, consultations lasted long before establishing the referent: *konwalijka dwulistna* ('May lily'), suspected from the beginning. If the context fails to prompt the right answer, in dozens of cases in which the same word functions as a possible name, especially when listed as such in specialist dictionaries, translators choose inadequately. Whereas 'liverwort' can stand both for *przylaszczka* and *wątrobowiec*, these two differ substantially. The first is a species of nice-looking early spring woodland flower, the second being the name of a specimen in a whole class of more primitive plants of the *Bryophyta* phylum. In the absence of Latin, if any information about their look, habitat (some liverworts are aquatic), reproduction techniques, geographical distribution or some other preferences is hinted at or included, the choice should be easy. The same applies to 'hemlock' –

---

[31] Available at http://plants.usda.gov/java/profile?symbol=CADI6 - 60k

choina – a tree and *szczwół plamisty*, a poisonous biennial of the carrot family.

When it comes to organisms that share more similarities, especially in sentences such as: 'In the X National Park you can see a range of interesting plants ... (a list of, fifteen or so follows)' it is easier to lose hope. The meadow vetchling can be *groszek łąkowy* (*Lathyrus pratensis, L. luteus*) or *groszek skrzydlasty* (*L. montanus, L. linifolius, L. macrorrhizus, Orobus tuberosus*), the meadow saxifrage – *koniopłoch łąkowy (Silaus pratensis, S.flavescens, Silaum silaus), skalnica ziarenkowata (Saxifraga granulata)* and *żebrzyca roczna (Seseli annuum, S.coloratum, S. purpureum)*. Although rose beetles are properly *kruszczyce*, sometimes they may be called *złotawce* (Durrell 1975b: 41 – Durrell 1999: 41). One of them, *Cetonia aurata*, is actually *kruszczyca złotawka*. But in this group there is also *złotawiec okazały* – *Potosia aeruginosa* (big rose chafer/beetle). Occasionally *kruszczyca złotawka* itself is called *złotawiec*. On top of it, *Euthystria brachyptera*, which represents a completely different order (*Orthoptera: Acrididae*) is referred to as *złotawiec*, too. The name seems to epitomise ambiguity and Latin may appear indispensable. On the other hand, several names of equal circulation frequency but not necessarily all known to a particular person and enumerated dutifully in succession in all texts may turn out to have one familiar referent: *żarnowiec miotlasty, janowiec miotlasty* and *szczodrzeniec miotlasty* are all *Sarothamnus scoparius* = *Genista scoparia* = *Cytisus scoparius*, in English 'broom' ('common broom', 'Scotch broom'). The presence of one Latin name, selected by international decision on the basis of chronology and usefulness (see Szujecki 1998: 30) would be desirable so as to compensate for a scarcity or unequivocality of contextual clues. It would satisfy human (and so translators') yearning for order and stability.

Meanwhile, scientific progress, which hinges on shaking old foundations, perpetually destabilises nomenclature. The very concept of species has been questioned as arbitrary and man-made. An eminent taxonomist, Ernst Mayr, rejects it as an opinion of 'armchair taxonomists', who, having studied no species and their populations in nature, drew conclusions that species of organisms were not concrete phenomena of nature (Mayr 1996: 262). Aware of the existence of hybridisation, allopolyploidy (having two complete sets of chromosomes derived from different species) and sibling species, Mayr nevertheless insists on the reality of the concept of species. His certainty stems from thorough analysis and his argumentation can be summarised, hopefully without distortion, as follows: a species is a species owing to isolating mechanisms, which do not always prevent interbreeding between two separate species. Despite the continuous introgression of alien genes and even the production of fertile offspring the two species consistently preserve essential integrity without leading to a fusion. In other words, no new species is ever created in this way. That, of course, does not rule out the opposite process of gradual speciation, in which two species are on their way to become one (semi-species) and they discriminate against introgressing genes of the other semi-species. In fact, it is agreed that a species is not a monolith. Firstly, there are subspecies reflecting geographical changeability, secondly, ecotypes formed as a consequence of adaptation to specific environmental conditions, and thirdly, an individual variation in appearance depending on such factors as age, time of the year, pollution, injuries, etc. *Trichogramma embryophagum* (*kruszynek leśny*, no English name), an insect parasite of other insects' eggs can serve as a pertinent illustration. Its two ecotypes do not differ morphologically but due to distinct mobility and ability to find shoots, needles or buds they have adapted to the penetration of a different environment, on which depends their effectiveness of attacking the eggs of the host. As a result, individuals which have grown up on the pine beauty's eggs (*strzygonia choinówka*) cannot be employed by foresters to fight the European pine shoot moth (*zwójka sosnóweczka*) (see Szujecki 1998: 26). On the other hand, genetics has contributed to discovering cryptic (sibling) species, betraying no or very slight morphological differences. Such discoveries lead to the multiplication of new species, for example, *Paramecium aurelia* was split into 14 in 1975 (Mayr 1996: 268). All in all, it must be noticed that whether an organism's status is that of a species, subspecies or race, they are usually assigned particular names, which are subsequently put into circulation.

New names may turn out to be those once sunk into oblivion and now brought back to life, not necessarily denoting the same plant or animal. For instance, the 'blackcap', which used to be called *pokrzewka czarnołbista*, has recently been officially endowed with the neologous name of *kapturka*; but the 'garden warbler', formerly *pokrzewka ogrodowa*, is now *gajówka*, an old name attributed in the past to various other *pokrzewkas* and to the subfamily of *Silvidae*, while the 'greenish warbler', once *świstunka zielonawa* is currently *wójcik,* which once belonged to *świstunka leśna* (wood warbler) (Strutyński 1972: 27-31) or referred to all *świstunkas* (Dobrowolski (ed.) 1991: 184). As demonstrated in Chapter Four, this nomenclatural inflation makes translators feel helpless. They at random give *gajówka* the meaning of 'any warbler', 'garden warbler' or 'any of the warblers until recently called *pokrzewka* plus a qualifier (e.g. *czarnołbista, cierniówka, jarzębata, ogrodowa, piegża*)'. Incidentally, the tendency to substitute longer names for short, single-element ones does not seem to be sufficiently justifiable since these often slow down locating a species within the broader taxonomic category. *Pokrzewka czarnołbista* was clearly one of *pokrzewkas*, *świstunka zielonawa* one of *świstunkas*, *brodziec leśny* (now *łęczak*), or *brodziec samotny* (now *samotnik*) were undoubtedly within a group of *brodźce*. The same cannot be automatically inferred from their new names.

Changes occur very fast. For instance, Tomiałojć and Stawarczyk (2003: 5-6) write that they follow the Polish nomenclature established in Mielczarek and Cichocki (1999), but admit they introduced updated versions accepted by international teams of experts. The two sources were published within a mere four years of each other. The main novelties are:

- a return to treating birds of prey as one order *Falconiformes* consisting of three suborders. The authors have replaced the old Polish name for *Falconiformes* – *drapieżne* with *szponiaste*, which lets reserve the former as a handy morpho-ecological term rather than an indication of the position in systematics;
- acceptance of granting a status of a species to clearly distinct geographical forms – hitherto subspecies.

Some of these alterations are induced by the advances of genetics. Gooders goes as far as to say that 'the widespread use of DNA as a key to ornithological, and other, relationships over the past few years has thrown systematics into disorder' (1998: 14). According to him, '[o]ne of the most significant results has been a multiplicity of 'new' species resulting from splitting previous species into two or more' (ibidem). Then he identifies three such splits that happened while the book was in production and predicts the inevitable recognition of further new species before it is published.

On the other hand, the differences in nomenclature may stem from the fact that authors of particular publications simply endorse different taxonomic systems. In the Polish edition of Jonsson's *Ptaki Europy* (1998) the author acknowledges he follows Voous; Dobrowolski in his *Ptaki Europy* (1991) does not declare adherence to a concrete system, but it is definitely not Voous'. Busse (ed.) (1991), however, promotes Wetmore's taxonomy, while Sokołowski's *Ptaki Polski*, published in 1992, already contains one almost like Voous' in Jonsson (1998). Table 2 shows the scale of variance by comparing only higher-rank taxa in but two of the sources:

Table 2: Comparison of names for bird families in two sources.

Latin name (family)	Dobrowolski 1991	Jonsson 1998	English name
Anatidae	Kaczkowate (9 subcategories)	Kaczkowate (5 subcategories)	Swans, geese, ducks
Scolopacidae	Słonkowate	Bekasowate	Snipe, sandpipers, phalaropes and allies
Charadriidae	Siewki	Siewkowate	Plovers and lapwings
Phalaropodidae	Płatkonogi	Included in Scolopacidae	Phalaropes
Recurvirostridae	Szablodziobowate	Szczudłonogi	Avocets
Accipitridae	Jastrzębiowate (10 subcategories), rybołowy belong to a subcategory of orłany (6 species)	Divided into 3 separate families: Jastrzębiowate, Rybołowy (Pandionidae) including 1 species, Sokołowate (Falconidae)	Vultures, eagles, kites, hawks, buzzards, osprey, harriers Falconidae - falcons
Phasianidae	Bażantowate	Kurowate	Pheasants, quails and partridges
Strigidae	Sowy właściwe	Puszczykowate	Typical owls
Sylviidae	Pokrzewki (6 subcategories)	Pokrzewkowate (no subcategories)	Warblers
Ploceidae	Wikłacze (includes wróble –sparrows)	-	Weaverbirds
Passeridae	-	Wróble	True sparrows
Emberizidae	Included in Fringillidae (łuszczaki)	Trznadlowate (separate family)	Buntings and American sparrows, Fringillidae – True finches
Panurus biarmicus (species)	Wąsatka – closest to Kurtodrozdy (Timaliidae) but also put among Paradoxornithidae	Wąsatka belongs to Kurtodrozdy	Kurtodrozdy – Babblers Wąsatka – bearded tit

Any modern language, but particularly English, native to so many people inhabiting different countries and spoken for international communication, has a problem of 'considerable overlap and confusion in names' (Gooders 1998: 13). For example, '*Charadrius alexandrinus [sieweczka morska]*, ... is known as Snowy Plover in North America, Kentish Plover in Europe, and Red-capped Dotterel in Australia' (ibidem). But unification of nomenclature within the English language is doomed to failure. Despite the existence of international bodies which make relevant recommendations, American names of species circulate in Europe, gladly embraced particularly by non-British writers and translators. Ridiculously, European birds which do not inhabit America are given American names on the strength of their close resemblance. The translation of *The Issa Valley* by Czesław Miłosz provides a salient example. Our *mysikróliki* become kinglets – American species *Regulus satrapa (mysikrólik złotogłowy)* and *Regulus calendula (ogniczek)*. The European *Regulus regulus*, is called 'goldcrest' in English. One does not have to stretch one's imagination too much to picture kinglets flitting from branch to branch in the pine crowns in Lithuania but they simply do not match the reality. When speaking English, naturalists from Poland, Scandinavia and Holland tend to refer to the elk (*Alces alces*) as 'moose' (both *łoś* in Polish), its American larger relation. North Americans need the word because 'the Rocky Mountain elk' stands for 'wapiti' (*Cervus canadensis*) and they also have the Roosevelt elk, both related to the red deer. The Europeans have no reason for substituting 'elk' for 'moose'.

Some regional or former alternative names have persisted for years after a concerted attempt to unify nomenclature. The ferruginous duck *(Aythya nyroca),* in Polish *podgorzałka*, for example, continues to be called white-eyed duck, although that has been reserved for *Aythya australis - podgorzałka australijska* . No wonder, if Webster's (1993) claims that white-eyed duck = white-eyed pochard refers to both species, and that *Aythya nyroca* is also called the hardhead. Likewise, *tchórz* instead of the canonical 'polecat' is sometimes presented as 'fitchew' or 'fitchet weasel', *goździk pyszny (Dianthus superbus)* as 'fringed pink', the name absent from Anioł-Kwiatkowska (2003) (henceforth AK), Blamey, Fitter, and Fitter (2003) (MB), Moore (1984) (PM) and Czekierda (1995), or *mieczyk dachówkowaty (Gladiolus imbricatus)* as 'sward lily', though of the above sources only Czekierda (1995) offers something similar: 'sword lily'. There are more examples of discrepancies between various sources in their indication of which existing name for a given species is most commonly used or preferred: *lilia złotogłów* – is 'martagon lily' or 'Turk's cup lily' (the latter does not figure in AK or PM, but in MB other possibilities are not mentioned except this one), but in the *Book of the British countryside* (1981) both are mentioned; *sadziec konopiasty*, in MB and PM called ‘hemp agrimony’, has this name as the third possibility in AK, *ostrożeń warzywny*, in MB and PM – the cabbage thistle, in AK this is the last of five possibilities. Sometimes we are assured that everything is sorted out, like in the case of ‘bulrush’. It was often attributed to *Schoenoplectus lacustris – oczeret jeziorny*, or formerly, *sitowie jeziorne*. According to (MB: 342), it was then officially awarded to *Typha latifolia*, hitherto alternatively called ‘reedmace’ and known in America as the ‘cattail’ – *pałka szerokolistna (wodna). Schoenoplectus lacustris* is now the common club-rush. In spite of the assurance, it suffices to take a look at the EUNIS inventory of habitats, where ‘reedmace beds’ under code D 5 prove that the decisions of specialists have an extremely limited influence.

To sum up, intralingual translation of categories denoting species takes place from one Latin name into another, from regional and obsolete native names into standard ones, from what some experts consider the standard into the standard valid for some others. Interlingual translation can begin when the translator manages to establish the source language referent behind the term, whose name may have been re-translated intralingually many times. Then the referent must be found in the thicket of target language categories, whose multiplicity results from continuous intralingual rewordings.

2.5. Folk and specialist categorisation – their mutual influence

Nomenclature used in classifications often borrows from non-specialist language. General language, constantly changing due to cultural processes such as urbanisation, is imprecise, prone to polysemy and uses words also functioning in professional contexts, although in a different meaning, or coins descriptive labels which successfully pretend to be specialist terms (e.g. total clearing instead of clear-cutting = *zrąb zupełny)*. As a result of urbanisation, words which were once part of the general lexicon become restricted to professional contexts. When rural inhabitants dominated the Polish population words such as *bronować* = to harrow, *biel* = fen carr consisting mainly of birch and pine or *pasternik* = common village pasture were widely understood, today’s young generation of city dwellers has usually no idea what these encode.

Certain specialist terms, when used in professional contexts in their everyday meaning, can cause serious problems. In an active conservation programme of in-forest mires in the Mazovian Landscape Park a removal of self-sown birch trees was proposed to stem the transformation of a valuable open landscape into woodland. Such practices are now accepted by nature conservation authorities, which have realised that passive protection has often proved counterproductive. Yet, in this case they reacted with suspicion and disbelief. That was due to one word *trzebież*. Instead of writing that it is needed to *wyciąć podrost brzozowy* (cut down birch growth), the proponents phrased it as *przeprowadzić trzebież brzozy* (to carry out the thinning of birch stand). In general Polish, as well as in the document under discussion, *trzebież* means 'cutting down'. In forestry terminology it signifies 'thinning of forest' to encourage a better growth of the trees left. No wonder the officials, probably accustomed to forestry jargon in similar projects, expressed reservations and demanded clarification. However, what disturbs communication most is the fact that general language 'terms' lack precision and people often have only a rough image/idea of a given thing/notion, whose referents vary regionally and whose correctness depends on individual experience steeped in a social and professional background. Specialists adopt such 'terms', attach a particular meaning to them, but non-specialists, reading about scientific issues in popular publications, tend to keep to their own, vague or mistaken understanding.

Complex relations between general language and scientific classifications with assigned terminology will be illustrated below by the words mire, bog, fen, swamp and their relatives. This overview follows Wheeler and Proctor (2000) and takes into consideration other sources, as indicated. 'Mire' has its heyday in specialist literature nowadays as an umbrella term for peat-forming peatlands. It was used in an even broader sense of natural and semi-natural peat communities but somewhat sparingly in older publications, for instance, by Sjörs, who explained that he used 'mire' for 'moss' ('bog') and 'fen' together, the term 'suggested by Godwin (1941: 260) and ... adopted by many Swedish ecologists when writing in English' (1950: 241). Sjörs admitted it was not a very common English word but the more common 'marsh' or 'swamp' 'have been used in quite different senses by British ecologists' (ibidem). Blamey, Fitter, and Fitter (2003) see 'mire' as 'a waterlogged ground – bog, fen or marsh' – even more broadly than Wheeler and Proctor (2000)[32]. 'Mire' 'has narrowed from its Old Norse origins, and usually implies deep dirt or mud on a track or a swampy place where one may become stuck; it is largely obsolete in everyday English'. Indeed, contrary to 'marsh', 'bog', 'fen' and 'swamp',

---

[32] See the end of section 2.3.

'mire' does not have its own entry in the *Book of the British countryside* (1981) nor is it mentioned as a habitat of any bird in Holden and Cleeves (2002) or Gooders (1998), where all the others occur in profusion.

Another name: 'moor' used to embrace all uncultivable wet peaty places – bogs, wet heaths and fens alike. Over the 20th century 'moor' acquired a more restricted meaning of 'bleak uncultivated upland, not necessarily peaty and often heather-covered'. The word 'moss', present in many place names, was gradually replaced by 'bog', a 16th century borrowing from Irish Gaelic. Today scientists agree that 'bog' denotes nutrient-poor, precipitation-fed mires/peatlands. 'Swamp' came to English as late as the 17th century 'as a purely descriptive word for a soft, wet spongy place (in American usage generally tree-clad)'. In science it refers to nutrient-rich peaty areas constantly or most of the year under water (reed swamps, alder swamps). It is in agreement with popular imagery: 'a low-lying area in which the soil is always covered with water, even in summer. Reeds are usually the only plants that grow in these conditions, though rushes sometimes occur. … It is usually 'confined to the edges of lakes, rivers and ponds.'(*Book of the British countryside* 1981: 452). Unlike 'moss', 'moor', 'fen', 'bog', 'marsh' and, to a lesser extent, 'mire'/'mere', 'swamp' does not figure in British place names – it is more traditional in former colonies or in the USA[33]. 'Morass's' arrival dates back to the same period. It is largely literary and equals 'mire' in its technical sense. 'Marsh' in common use is just a wet, muddy area. For scientists and naturalists[34] it is confined since Tansley's *The British Islands and their vegetation* (1939) to the wet vegetated places in which 'peat is not formed generally and in quantity'. 'Fen' is viewed in Britain as 'a waterlogged place with reeds and other plants growing in the patches of standing water', the underlying peat 'is saturated with lime' (*Book of the British countryside* 1981: 174, Blamey, Fitter, and Fitter 2003: 17). Scientists use 'fen' in a similar sense but they require the peat to be alkaline and nutrient-rich, not necessarily lime-saturated. A fen overgrown with reed is considered degraded, the sedge-moss carpet being most desirable. As can be seen, English scientific names come from everyday language but do not always coincide with the current popular understanding.

Popular understanding is further characterised by simplification, which here boils down to blurring the distinctions between types of mires. The

---

[33] See the list of UK Special Areas of Conservation (Natura 2000) or Ramsar sites at http://www.jncc.gov.uk

[34] See e.g. the *Book of the British countryside* (1981: 66, 174, 280-1, 452), Blamey, Fitter, and Fitter (2003: 17).

comparison of three monolingual dictionary definitions should be quite indicative:

Table 3: Definitions of types of mires in three general English dictionaries.

	Collins 1995	OALD 1989	Webster's 1993
Bog	An area of land which is very wet and muddy	Wet spongy ground formed of decaying vegetation	Wet spongy ground where a heavy body is likely to sink: Quagmire, Morass; *esp*: an inadequately drained area rich in plant residues, usu. acid in reaction, frequently surrounding a body of open water, and having a characteristic flora (as of sedges, heaths and sphagnum)
Fen	An area of low, flat wet land, especially in the east of England	Area of low marshy land	Low peaty land covered wholly or partly with water unless artificially drained
Marsh	Wet, muddy area of land	Low-lying wet land	A tract of soft wetland: Fen, Swamp, Morass; *specif*: such a tract of land often periodically inundated and treeless and usu. characterized by grasses, cattails, or other monocotyledons
Mire	1. (Figurative use) 2. Dirt or mud, literary use	Swampy ground or bog; soft deep mud	Wet spongy earth: Marsh, Swamp, Bog
Morass	Only figurative use	Stretch of low soft wet land; marsh	A tract of soft, swampy, or boggy ground: Marsh, Swamp
Swamp	An area of very wet land with wild plants growing in it	Soft wet land; marsh	Wet spongy land saturated and sometimes partially or intermittently covered with water: water-logged imperfectly drained land unsuitable for agriculture without artificial drainage, *esp*: such land supporting a natural vegetation predominantly of shrubs and trees and often intergrading into grassy marsh on the one hand and wet forest on the other

The table clearly shows how little variation is registered in the meaning of the six words. Even the bulky Webster's, which prides itself on precision and comprehensiveness, projects an impression that there is not much difference between them. In the case of 'bog' and 'swamp' it approximates the specialist meaning but stops short of the most important features. Remarkably, this evolves towards the resemblance of the Polish lexicon, where there is one all-encompassing term *bagno*. Other existing words are descriptive – sometimes constrained in a self-evident way, sometimes flexible.

In Polish the situation is clearer. Basic scientific terms sound scientific (*torfowisko niskie, wysokie, przejściowe, zasadowe*, etc.) and cannot be confused with everyday usage. If someone does not feel like discriminating between types of mires, they have at their disposal a convenient umbrella term. If, in turn, somebody wishes to sound more professional or literary, a range of possibilities is waiting: *bagnisko, błoto/a, moczar, trzęsawisko, grzęzawisko, mokradło, oparzelisko, or rojst*[35]. Admittingly, some more detailed specialist terminology, as borrowed from general Polish, might be counterintuitive in individual cases (*mszar* vs. *mechowisko*), but it is a rather marginal problem. Problems emerge when it comes to translation. Bilingual dictionaries simply list possibilities selected in an arbitrary fashion without any guidance as to the context a given word fits or, perhaps more importantly, does not fit. As demonstrated above, monolingual dictionaries rarely help, so consultations with specialists or relevant up-to-date literature become indispensable. In order to provide evidence that consulting bilingual dictionaries can only initiate a drawn-out process of tracking down the meaning six dictionaries are compared – three Polish-English and three English-Polish ones.

[35] The word *rojst (rojścik)* is of Lithuanian origin – *raistas – mszary, mokradła, torfowiska bagienne, torfianki* (Kopaliński 1983). In English, the words given by Kopaliński as explanation stand for: raised bogs, wetlands, peatlands, peat cuttings-over. It was used in the former north-eastern part of Poland, today belonging to the independent states of Lithuania and Belorussia. It is immortalised in the titles of literary works such as *Bunt Rojstów* by Józef Mackiewicz or *Rojsty* by Tadeusz Konwicki. However, it is not a term well-known to contemporary Poles.

Table 4: English equivalents of types of mires in three Polish-English dictionaries.

	Czekierda 1992	Stanisławski 1989	Linde-Usiekniewicz 2004
Bagno	Bog, swamp, fen, morass, quag (mire) + five types e.g. *porośnięte trzciną* – reed swamp	Bog, marsh, swamp, mire, fen, morass, quag, slough	Bog, marsh
Błota	Swamps, marshland, marshes	Swamp, swamps, marshland, marshes, bogs	Marsh, marshland
Grzęzawisko	Quag, quagmire	Quag, quagmire, literary: morass	Quagmire
Mechowisko	-	Moss-grown spot	-
Moczar	Bog, marsh, swamp, fen, morass	Bog, marsh, swamp, literary: morass	Bog, swamp
Mokradło	Marshy ground, bog, swamp, marsh	Marshy ground, swamp, bog, marsh, morass	Bog, marsh
Mszar	-	1) Bot. variety of moss 2) (Mokradło) morass	-
Oparzelisko	-	Morass, bog	-
Rojst	-	-	-
Trzęsawisko	Swamp, quaking bog, quagmire, slough, hag	Bog, swamp, quagmire, slough	Swamp, bog

Table 5: Polish equivalents of types of mires in three English-Polish dictionaries.

	Czekierda 1995	Stanisławski 1990	Linde-Usiekniewicz 2002
Bog	Hydr. bagno, moczar, mokradło; bot. torfowisko	Bagno, moczary, trzęsawisko	1.(Marshy ground) bagno, mokradło, trzęsawisko 2. Torfowisko
Fen	Bagno, moczar, fenland – teren bagnisty	Bagnisko, moczar, zalewisko	Mokradło
Hag	Trzęsawisko	Trzęsawisko	-
Marsh	Bagno, moczar, mokradło, obszar podmokły	Bagno, błoto/a, moczar, moczary	(Terrain) bagno, moczar (region) bagna, moczary
Marshes	Błota (geogr.) + 3 types + marshland – żuławy, błota, obszar zabagniony, teren bagnisty	-	-
Mire	-	Błoto, bagno, muł	Literary: bagno, grzęzawisko
Moor	Wrzosowisko, torfowisko	Wrzosowisko, torfowisko	Wrzosowisko
Moorland	Wrzosowisko, torfowisko	Wrzosowisko, torfowisko	Wrzosowisko
Morass	Bagno, moczar, trzęsawisko	Bagno, bagnisko, moczar, moczary, grzęzawisko, mokradło/a	Bagno, grzęzawisko
Moss	Not defined as a habitat at all	Torfowisko	-
Quag	Grzęzawisko	Bagno, trzęsawisko	-
Quagmire	Bagno, trzęsawisko	Bagno, trzęsawisko	Bagno, grzęzawisko, trzęsawisko
Swamp	Bagno, trzęsawisko, moczar, obszar podmokły + 3 types, swamps - błota	Bagno, błota, moczary	Bagno, trzęsawisko, moczary

Apart from the high context-dependence of the enumerated words, several other comments as to complications in usage must be made. The word 'wetland' does not appear in the dictionaries as an entry at all, although it is the best equivalent of *mokradło, obszar/teren podmokły. Trzęsawisko* has its scientific translation as 'quaking bog', but as already pointed out, it is rather unjustifiable to call all forms of *trzęsawisko* 'bog', hence 'quaking mire' would be more appropriate. Some terms used by the same author in the English-Polish version do not turn up in the reverse version (e.g. 'fen' in Linde-Usiekniewicz 2004). Some others are attributed surprising equivalents e.g. fen – *zalewisko* (floodland), moss – *torfowisko* (why not *bagno*, as elsewhere, or *torfowisko wysokie* to tell the whole truth not half of it?), *mszar* – is roughly *torfowisko wysokie*, so 'bog' and not 'morass' should be offered as an equivalent. It is difficult to understand why 'bog' is said to be *torfowisko* and 'fen' is not, although both develop on peat, especially as at the same time 'moor', which may or may not be peaty, has been given *torfowisko* as a possible translation. Lack of *torfowisko* among Polish equivalents of 'mire' is a real shame[36]. *Oparzelisko* is not 'bog' but a particular spot in a 'fen' which never freezes in winter probably because of spring waters gushing out there. As a word which nowadays has a literary flavour it can be rendered as 'morass' but with sufficient qualification: 'never-freezing morass' or 'ever-steaming morass'.

Having presented the interaction between folk and specialist categorisation on the basis of mire terminology, let us turn to some informal 'semi-scientific' categories denoting birds, taking 'warblers' as an example. The *Book of the British countryside* (1981: 487-88) divides them into three main groupings: leaf, scrub and marsh warblers. Gooders adds 'tree warblers', which do not occur in Britain (1998: 229). In Polish, this informal categorisation cuts across slightly different lines: 'marsh warblers' corresponds to two groups: *świerszczaki* and *trzciniaki*, the rest correlating as follows: tree warblers – *zaganiacze*, leaf warblers – *świstunki*, scrub warblers – *pokrzewki* (Dobrowolski (ed.) 1991). The English phrases reflect an attempt to epitomise the groups' ecology. The Polish ones do not generalise in this way, are slightly artificial and not customarily presented in guidebooks. They simply group birds according to their names, which unlike in English, do not all comprise 'warbler'. In several cases the species whose names do not contain the 'brand' label are thrown in as they share significant ecological features: e.g. *brzęczka*

[36] See e.g. Grootjans and Wołejko (eds) (2007) for the most up-to-date text with Polish and English versions placed parallel to each other for easy comparison.

(Savi's warbler) and *strumieniówka* (river warbler) are put together with a number of species called *świerszczak* (grasshopper warbler), *trzcinniczek, rokitniczka, wodniczka* or *łozówka* with those called *trzciniak*, while *pierwiosnek* and *piecuszek* with *świstunka (leśna, zielonawa, żółtawa*, etc.). All scrub warblers used to be *pokrzewka* + qualifier in Polish. Recently, this uniformity has been disrupted and now, as illustrated above[37], there are a few species without *pokrzewka* in their names. The lack of straightforward correspondences leads to awkward situations in translation – if one writes that marsh warblers are *świerszczaki i trzciniaki*, Polish readers, unaccustomed to this kind of categorisation, may conclude that the group consists of just two species *świerszczak* (grasshopper warbler) and *trzciniak* (great reed warbler).

Among other semi-formal or both formal and informal categories worth elaborating on are: undergrowth, brown mosses (*mchy brunatne*), wildfowl, water birds, water fowl, grouse, finches, chats, buntings, redstarts, waders/shorebirds, swallows, sandpipers, sawbills/*tracze*, dormice, catchflies, chickweeds, all kinds of cultivars of fruits, pyrolas, copse/coppice/grove, fledgelings, dry ground forests, deer, midges, lacewings, gnats/mosquitoes/*komary*, *ważki*, *wodno-błotny*, *gołębie*, *płochacze*, *jagody, borówki*, *tobołki*, *orchidee/storczyki*, *szuwary*, *sitowie*, *zarośla, chaszcze, turzycowiska, wrzosowiska*, *bór, grąd*, *starorzecza, bagna, mateczniki, ostępy, ostoje, uroczyska, rogi, dziuplaki, złotawce* and so forth. Born in the minds of people representing different cultures, embedded in a wide range of natural conditions, they escape easy transfer from one language into another. They are categories whose boundaries are drawn differently by the users of the Polish and English languages. The translator is compelled to identify an individual concept behind each verbal representation. In many cases it is not just the 'mean' concept in the mind of the average member of society – the average participant in one or the other culture, it is the concept behind the specific use, referring to a particular instance of *grąd, wrzosowisko* or warbler. Each conceptual category is built around a prototype, the indication of which is a moot point within one language community. Controversies over prototypes entail diverse opinions concerning the extent to which other category members are peripheral. Sometimes categories with clear-cut demarcation lines in the minds of most people who have been taught their specialist definitions at school become blurred due to the constant tension between naïve and expert understandings (cf. Taylor 1995: 73).

---

[37] See pages 60-61.

As languages behave in a more conservative way than non-verbal concepts (cf. Maćkiewicz 2004b: 195), rapid changes in expert understandings cause a hiatus between lexical items in circulation via dictionaries and handbooks and new expert-generated categories with their verbalised labels. Prototypical effects yielded by categorisation: background framing, fuzziness, gradability and stretchiness, whose existence has been experimentally confirmed by cognitivists, are a curse and a blessing of translation. They are a curse because the task of languages and translations is to put the chaotic world into order. If the way humans cut up this world is so imprecise and flexible, this task seems undoable. However, anything 'undoable' has always fascinated and spurred people to extra effort and to find appropriate solutions. In this sense they are a blessing. The search for solutions to translation problems with most of the categories listed at the beginning of the previous paragraph is launched in the next chapter, which intends to show the influence the complexities of classification laid out above exert on translation practice, and to link them with questions of cultural determinism.

2.6. Recapitulation

This chapter has emphasised that fuzzy boundaries which tend to shift (stretchability), gradability of membership and dependence of membership on the classifier (background framing) typify both folk and specialist categories. Multiple intralingual translations of the same verbalised category between disciplines, schools of thought, language varieties, registers and contexts give rise to terminological chaos. The translator may have to go through yet another process of intralingual translation in both source and target languages in order to match concepts and their verbal representations in each of them. For example, encountering the above mentioned obscure, regional term *majownik dwulistny*, s/he translates it into the standard *konwalijka dwulistna* before being able to render it interlingually as 'May lily'. Even if such operations are not necessary, the translator must still be aware of the extralinguistic context. Its recognition is crucial for making appropriate choices in situations such as: Does one need a forestry or conservationist version of terminology?; Is *turzycowisko* 'tall sedge swamp' or just 'sedgebed'?, and many others discussed above.

CHAPTER THREE

THE CATEGORISATION OF NATURE IN POLISH AND ENGLISH TEXTS FOR NON-SPECIALISTS

3.1. Perspective

The third chapter is motivated by two forces struggling within humans: indifference to the overcomplex natural world and a desire for harnessing it through neat classification once and for all. Translators must resist both. The following dialogue from *Daniel Martin*, by John Fowles illustrates why they should resist the first temptation:

> "Why isn't it enough that I just love it here? That I don't want to know all the names and the frightfully scientific words."
> "Because you shouldn't justify contempt from ignorance. In anything."
> (1978: 350)

On the other hand, John Fowles' books show that nature, human nature included, escapes orderly, finite categorisation. In *The French lieutenant's woman* (1996: 234) the narrator observes that today we take '[the] priority of ecology over classification' for granted, unlike a century or two ago. In other words, the world is a continuum – a tangled web of entities, which can be singled out in a variety of ways. Attempts to classify the environment as well as human behaviour according to frames designed *a priori* are proved to fail repeatedly. Translators cannot be like Anthony from *Daniel Martin*, who

> had a much more genuine hatred, almost a fear, of what could not be collected, classed, precisely defined, noted down: I mentioned earlier his obsession with the Dactylorchis group of British wild orchids, which hybridize with bewildering frequency – I think the fluid frontiers between their species seriously upset him, and they were rather less a botanical challenge than a nagging flaw in his would-be highly ordered nature of things. (1978: 361)

Fluid frontiers of categories cannot be treated as a nagging flaw but as a challenge to investigate perspectives from which a particular category is drawn.

Perspective is understood here as a parameter of Langacker's notion of imagery. It consists of four factors: orientation, vantage point, directionality and subjectification. Orientation refers to how the conceptualiser is situated in physical space while watching a specific scene. Vantage point is the point from which he or she does it – from inside or from outside. Directionality concerns the direction of his or her eye movement (e.g. from top to bottom). Subjectification is about expressing one's own or somebody (something) else's point of view (Tabakowska 1993: 64-69). In folk categories, which have been adopted by scientists and experts-practitioners, what changes first of all is the vantage point and subjectification (Maćkiewicz 1996: 248). The observer, so far situated inside the described fragment of reality, moves outside and tries to separate facts from opinions, comprehensive cognition from short-term utilitarianism. Across related disciplines orientation and directionality may shift too. The forester looks at the trees and the herb layer from a standing position, the phytosociologist – kneeling on the ground and examining the forest floor carefully. The former moves his eyes from the trees to the bottom, the latter always starts with the bottom, assigning the upper layer the role of an auxiliary indicator of habitat type. Apart from perspective, three other dimensions of imagery are advised to catch the translator's attention: comparison that enables the measuring of the level of conventionalisation of prospectively equivalent verbalised categories, selection that lets us determine the degree of salience and abstraction, due to which the degree of schematicity can be established. In particular scenes their analysis reveals if, how, and where verbalised categories of two languages match and where the respective conceptualising systems leave gaps which can only be bridged by circumlocutions, explanatory notes or other cumbersome methods.

3.2. The environmental and cultural background

In the case of nature terminology imagery is closely connected with the environment the two languages and cultures have evolved in. What kind of environment has moulded folk terminology of the English language is a question the answer to which begs another question: Which language variety do we mean? Polish is confined to one area in central Europe, English spreads over continents. I concentrate on British culture, primarily because all other Anglo-Saxon cultures stem from it and have preserved some of its traits. Secondly, the UK and Poland both lie in Europe, which makes it easier to compare natural worlds encapsulated in Polish and English lexical items. Both countries belong to the European Union, which necessitates looking for similarities and common solutions to the problems of nature conservation. Occasionally American English categories are addressed, too, as they circulate in translations from Polish.

The environments of Great Britain and Poland have been shaped by very different geographical and socio-political conditions. The former is an island, boasting a multitude of marine habitats and inland habitats thriving on a mild but windy and humid climate. Islands as a rule are poorer in species than mainlands, where free migration is easier and the effective extermination of a species takes up much more time. No wonder, therefore, that with the comparable level of hatred to animals with which humans compete for food and other resources on the Continent and on the island, the wild boar, elk, lynx, wolf, beaver, not to mention the bear and the European bison, though still surviving in continental Europe, in Britain are the song of a remote past. Natural isolation has affected the course of history. While the Continent has been plagued by on-and-off wars, Britain has remained relatively peaceful and safe. People could focus on the development of agriculture and industry. As a result, the population was stable and since the 18th century has kept growing and voraciously exploiting the natural resources, most remarkably denuding the landscape of tree cover. As early as 1350 woodland (then almost entirely broadleaved) constituted only ten per cent of the total area (Evans 1992: 13), by 1919 dropping to three per cent (Evans 1992: 60) to rise again to ten per cent in the 1990s and twelve per cent in 2008, but chiefly due to artificial plantations of non-native conifers[38]. Devoid of trees, large tracts of land became exposed to soil erosion in the moist climate, which together with the immense popularity of sheep breeding prevented the regeneration of forests. Famous heaths and heather-clad moorlands emerged instead, in places unsuitable for pastureland and have since been maintained as hunting grounds. They have been managed to support the profit-generating game species: the ling-eating red grouse and the red deer. Hunting for grouse on moorland is one of the tokens of British culture.

Extensive open landscapes have always been favoured. The supreme instantiations of this national preference are the Enclosures of the 16th to 18th centuries, the Industrial Revolution and its aftermath and the post World War II intensive agriculture boom. The Enclosures consisted in cutting private enclosed parcels out from the land used communally by a

[38] See http://www.forestry.gov.uk/statistics (accessed 18. Dec. 2008).

village and combining small scattered allotments into vast one-piece properties. One of the reasons was the creation of uninterrupted pastureland for sheep, whose wool was in high demand from the flourishing textile industry. For the environment there were two major consequences – one immediate and direct, the other materialising in the long run. The first was the replacement of a chequered pattern of small fields with large ones fenced in by dry-stone walls or hedgerows. The second was the removal of copyholders – peasant tenants – from their land, which generated legions of idle hands, the dispossessed vagabonds. Later, however, it appeared to be one of the factors contributing to the Industrial Revolution. This period in British history further intensified the loss of tree cover and animal species whose habitats were heavily transformed and polluted. This was coupled with a craze for collecting the eggs of wild birds, never pushed to such extremes in Poland. The intensive farming methods of the 20th and 21st centuries have led to the large-scale uprooting of hedgerows and mid-field trees to make room for larger fields and the turning of pastureland, heathland, wetland and wasteland into arable land. Little natural ponds or depressions filled with mires have been drained and levelled. The decline of open landscape species of birds and amphibians continues unabated. In a nutshell, the environment of Britain is not natural at all because of long-lasting prosperity manifested in intensive economic activity and overcrowding. Native woodland is scarce, landscape typically open. Access to land due to the law of trespass and ownership structure (little public land, the rest owned by the very few) is extremely limited. Thus society has long been predominantly urban, very sentimental about the natural and the wild, which, however, they want to see tamed and comfortable to experience. The feeling of disgust to anything untamed, irregular, haphazard, untrimmed and rough, though not unique to the British, has imprinted itself on the British landscape with exceptional ardour, the effect exacerbated by the nation's entrepreneurial skills. This is how it is described by John Fowles in *A maggot*:

> The period had no sympathy with unregulated or premordial nature. It was aggressive wilderness, an ugly and all-invasive reminder of the Fall, of man's eternal exile from the Garden of Eden; and particularly aggressive, to a nation of profit-haunted puritans, on the threshold of an age of commerce, in its flagrant uselessness. … even its natural sciences, such as botany, though by now long founded, remained essentially hostile to wild nature, seeing it only as something to be tamed, classified, utilized, exploited. (1985: 15)

Although the above passage refers to the 18th century, similar statements can be found elsewhere about earlier and later periods (e.g. Evans 1992: 17-18).

To return to the question of geographical location, in contrast to Britain, Poland's links with the sea are much looser. At 498 km its coastline is far from the shortest among European states but in the past its ties with the Baltic were recurringly severed due to invasions by stronger foreign powers. Situated in the centre of Europe, Poland is partially influenced by the marine and partially by the drier continental climate, which brings cold in winter and heat in summer. Numerous species have either the western or eastern, southern or northern limit of their range of distribution here, which definitely fosters biodiversity. In the past the location was rather unfortunate for Poles as a nation. Surrounded by contending empires they were often oppressed, their abodes and economy ravaged. In a sense the people's misfortunes were favourable for wildlife. Depopulation and hardships of over a century under enemy occupation hampered economic development and lessened pressure on natural habitats. While to the west of the Elba river peasantry was relieved of serfdom by the 16th century, to the east of it its most slave-like form survived till the second half of the 19th century. This hindered the process of transformation of agriculture into industrial enterprises, which was well underway in Britain. As early as the 18th century '[a]lthough demographically still predominantly a rural nation, it was one in which virtually all vestiges of a peasant economy had disappeared' (Bunce 1994: 8) and '[b]y the middle of the nineteenth century Britain had become a thoroughly industrial nation, leading the world in manufacturing and trade.' (ibid.: 9).

Even after the emancipation of peasants and granting them land ownership, farmland in Poland remained wildlife-friendly because primitive cultivation methods continued as before and instead of large farms, small allotments prevailed. Refuges for animals, such as clumps of trees, herb-rich balks, little ponds and marshes, etc. interspersed the mosaic of arable fields, flowery meadows, pastures and orchards. Poverty, underinvestment and the strong feeling of bond with one's belatedly given land helped this type of countryside to survive the 20th century turbulences, including the communist rule and its vision of large collective state-run farms.

For over twenty years the agricultural structure in Poland has been changing for the worse from the nature conservationist's point of view for three reasons: the intensification of production, the cessation of extensive management of least productive wet and dry grasslands and the loss of land to built development. Despite such unfavourable conditions, the fate of many species of fauna and flora, which in western Europe have either vanished, are on the verge of extinction or are declining rapidly, comparatively, is fairly good. Polish forests, too, have done better than those in the UK. Not only is there much more virgin or semi-natural woodland but the commercial forests have not been managed as mere plantations but as home for wildlife as well. At the end of the Middle Ages, Poland had four times as much woodland area as Britain and in the 18th century, when the latter was almost denuded, between 31 and 43 per cent of the total surface area (Zaręba 1986: 6-7). According to GUS (Central Statistical Office) currently it is 29.1 per cent (31 Dec. 2009). Unlike the British scanty woods, Polish forests still harbour large carnivors and herbivors. Worryingly, though, the knowledge of botany and zoology, which Polish city and country dwellers alike display is extremely poor and their ecological awareness seems lower than among the British, who have learnt to appreciate what they have lost or have to go to great lengths to retain (but admittingly lag behind the Dutch or the Swiss). Paradoxically, such ignorance, by keeping people away from wildlife habitats, has counterbalanced the lack of law of trespass in Poland. Until recently at least, disturbance, trampling and overall deterioration of living conditions for plants and animals had not been as widespread as in the case in which freedom of roaming had been exercised more eagerly by bird watchers, egg collectors (this has never been as popular as in Britain, incidentally), butterfly catchers, trekkers, and the like.

3.3. Verbalised categories – dimensions of imagery

The above sketchy comparison from the historical and contemporary perspective of an environmental and cultural background, in which Polish and English nature terminology are immersed, is intended as a reference point for a more detailed examination of images behind specific terms. The terms come from numerous sources such as albums, guidebooks, maps, brochures, reports as well as narratives translated from Polish into English, or vice versa. They may be written by specialists, but not for specialists. Some of them are not extracted from translated texts, but since they conveniently illustrate problems akin to those that crop up in translations, they are put on an equal footing. They lend themselves to the discussion on the shifts in various dimensions of imagery across language borders particularly well because they do not attract attention as errors, which are targeted in the next chapter. In this section the largest quantity of potential interpretations of selected terms is analysed, not necessarily with reference to specific uses in texts. The terms, which in themselves imply a lexico-semantic field under scrutiny, appear in alphabetical order, first those that cause most trouble in translations from Polish, later those that seem more difficult or equally challenging in translations from English. The scale of complication is assessed on the basis of my own experience and comparison of the texts enlisted in the bibliography. As any evaluation, it is subjective and not always defensible on second thought but necessary for organisational purposes. The next section foregrounds the behaviour of categories in concrete contexts in four situations: when a term is obscure to most contemporary readers in the source language, when slight shifts in register evoke disparate though suitable images, when puns are involved and finally, in the case of onomatopoeia. Both sections pick out material from a thorough documentation collected for this book.

Bagno

As already observed in Chapter Two the eight English equivalents at our disposal: bog, fen, marsh, mire, moor, moss, morass and swamp make but an illusion of choice. In modern everyday usage, only at the level of perspective a more consistent categorial division can be noticed: morass and mire are set apart as literary and figurative, moss as an obsolete predecessor of 'bog', whereas moor as fundamentally belonging to a different lexico-semantic field overlapping with that of *bagno* in a limited number of individual occurrences (see *Wrzosowisko* beneath). Bog, fen, marsh and swamp are highly schematic. Scientific distinctions might influence the relative salience of a given word in a particular context in individual cases. Generally, interpretations vary – what is profiled (designated) by each of the words pivots on the conceptualiser's cognitive environment. Swamp and bog are in all likelihood seen as concealing underwater fathomless liquid mud deposits, hence as more treacherous than marsh. Fen tends to be associated with peaty ponds fringed with reeds – a post-exploitation landscape, or with peaty meadows. However, immortalised in many place-names, it has lost its ties with any real phenomenon of the world. Instead, when people hear 'fen' they imagine the spot called so, though it may bear little or no resemblance to what it originally was. Marsh has the lowest degree of specification. Entrenched in proper names (often as 'Marshes') it sometimes denotes areas underlain with peat such as The Biebrza Marshes or The Prypet Marshes, which stands at odds with the scientific definition. Cases of this kind excepted, in this work I never tire of insisting that the decision which English word is picked out ought to be motivated by up-to-date specialist uses. Even narratives provide clues. In Miłosz (1989: 200) 'bagnu pełnemu ciemnych badyli' is correctly rendered as 'the dark-weeded swamp' (Miłosz 1981: 221). Any other choice except 'fen' would ignore the clue offered later in the description that the site is 'a hollow thick with alder'.

Bagno also acts as an abbreviated version of *bagno zwyczajne (Ledum palustre),* in English the Labrador tea[39], a small bush met in bogs and bog woodlands. The images behind the two *bagnos* seem to rule out the slightest chance of confusion and yet, in Pudlis (2003:18) 'Stanowisko bagna zwyczajnego w rezerwacie "Czarny Koń' becomes in English 'A common swamp in the Black Horse reserve'. The word *stanowisko* – 'site', 'locality' (of a plant) already narrows down the scope. Next comes the modifier *zwyczajnego* (common) put after the noun, as customary for scientific names of species, not before, as in other collocations. An automatic association of *bagno* with a boggy area wins. The translator has equalled the profile of *Ledum palustre* with elements of the maximal scope for it, which themselves are the profile of bog (woodland). The most schematic image forces out the specific instantiation. On top of it, the Labrador tea does not grow in swamps – a habitat too rich in nutrients. One has to overstretch one's imagination to visualise the plant among reeds or alders.

Barć/ bartnictwo

In Polish, there is a crucial distinction between 'drzewa z gniazdem pszczół' – 'trees with bee nests' and 'drzewa bartne' = 'trees with holes made by man to attract bees and managed for bee products'. The hole with a bee colony is called *barć*. The English language does not have a term to denote the latter and sometimes uses the corruption of the Polish word *barć* – barc (cf. *The praise of honey* 2005). Beekeeping in a country which dispensed with their largest woodlands a long time ago cannot be associated with barcs, which can only be made in thick trunks of old but healthy trees, capable of coping with such an assault on their health. Although now defunct, this kind of beekeeping (*bartnictwo*) survived until relatively recently and is part of Polish culture, in which the occupation enjoyed high esteem and privileges. The abundance of surnames and placenames reminiscent of *bartnictwo* bears witness to the past popularity and importance of the activity.

---

[39] Blamey, Fitter, and Fitter (2003: 108) give this name as the only officially recognised, but the marsh tea is in widespread use in publications.

Precisely because of a cultural gap in respect of beekeeping traditions between Poland and Britain translation poses a problem. First of all, the levels of lexical conventionalisation differ strikingly. What in Polish is expressed via a single lexeme, in English it is expressed through circumlocutions:

bartnictwo – primitive beekeeping (Rysiewicz 2000: 17/18);
barć (initially referred to the natural tree hollow with bees, later also to the trees including the hollow made for bees) – bee tree (ibid.: 14), primitive beehive (17), tree hive (93);
sosna bartna (a photograph caption) – pine used as a hive for wild forest bees (ibid.: 70).

Secondly, levels of schematicity do not match well. A Pole automatically imagines *bartnictwo* with its attribute – barcs and simultaneously juxtaposes it against *pszczelarstwo* (beekeeping) in its traditional and modern manifestations. Confronted with 'primitive beekeeping' a Brit retrieves from his or her cognitive base a gestalt, presumably without barcs even in the background. After all, 'primitive' may refer to the inefficient construction of man-made beehives and tools at beemen's disposal. If the translator opts for foreignisation and writes 'barc' and an explanation, there is a chance that the reader will visualise roughly the prototypical thing. With 'primitive beehive' such hopes are dashed, 'tree hive' lending itself to inadequate interpretations, including a familiar wooden box hoisted on to a tree branch.

Borówki/Jagody

From the scientific point of view there are three species of *borówka*: *brusznica*, *czernica* and *bagienna* – the cowberry, bilberry, bog bilberry. Folk usage varies – in some regions *borówka,* metonymically, stands for *borówka brusznica*, in some others for *czernica*, *bagienna* bearing the name of *łochynia* . In English the list of plants whose names contain *-berry* is impressive, thus translating *borówki* into 'berries' increases the number of possible images dramatically. *Jagody*, in turn, in professional jargons of biologists, horticulturalists, etc. refers to all sorts of berry fruits. Colloquially, everywhere in Poland but in the south the word denotes the bilberry. It is often extended over the bog bilberry, which is much less common. In Britain the names for bilberry differ across regions, the official one coming from northern England, shifting into whortleberry in the south and blaeberry in Scotland (Blamey, Fitter, and Fitter 2003: 106).

In both languages the use of particular names hinges on the perspective of the speaker: expert versus non-expert, southerner versus northerner, etc. and on the immediate context. If a Pole says in July or August: 'idę dziś na jagody' (I'm going to pick berries today), for the northerners the message is clear. For the southerner it is nearly as obscure as the same utterance in English. Certainly, the person does not intend to work in the garden, so only wild berries count. If in the vicinity bilberries grow in abundance, whereas wild strawberries, and crowberries not at all, plus one knows that cowberries and cranberries mature later, they may succeed at guessing. Salience is relative to the circumstances. Likewise, a higher degree of specification is required when clues may not be accessible to the interlocutor unless the sender of the message wants to refrain from elucidation.

Bór/las/puszcza

In the Polish lexico-semantic field of general labels for a sizeable area of tree stands with a more or less developed undergrowth, there are three lexemes: *las, bór* and *puszcza*. In addition, a large number of two-word combinations appears, but they are not of interest here. In English, the corresponding field contains wood, forest and woodland as single element items and coniferous (needleleaf) forest/wood/woodland to stand for *bór*, primeval/virgin forest and ancient woodland to match *puszcza*. As can be seen, the English has not got single item words for what the Polish language has. The latter has also two word equivalents of virgin forest – *las dziewiczy*, primeval forest – *las pierwotny*, needleleaf forest – *las iglasty/las szpilkowy.*

Alongside the specialist definition of *bór* as coniferous forest a folk understanding of it as 'bigger woodland' exists (Marczewska 2002: 55). This fosters mistranslation as translators all too often rely on the folk terminology, disregarding the context. Despite educational efforts, the conventionalisation of *bór* as vast and dark woodland is so firmly established that acquired knowledge languishes in the remotest corners of Polish minds, and *las iglasty,* though not a short, convenient label, takes over from *bór.* Saying *las iglasty,* most Poles do not consciously enhance the salience[40] of the notions [LAS] and [IGLASTY] but profile a different entity from *bór.* Interestingly, one expert term has been determinologised and the terminologisation of the other has been stubbornly refused recognition. Incidentally, *las szpilkowy* continues to function almost

[40] As suggested by Langacker for pairs such as pork: pig meat (1991: 62).

exclusively as a specialist term, so the perspective matters here. *Puszcza* derives from *pusty* – empty, uninhabited and although prototypically 'ancient extensive woodland', it lingers in the names of Polish forests seriously contracted in area and altered in structure. In translation, it should become simply 'forest', perhaps with one exception of Puszcza Białowieska, traditionally and deservedly rendered as 'primeval forest'. *Las dziewiczy* and *las pierwotny* are reserved for vast forests which have never been managed and have preserved their natural qualities. Remarkably, names of such forests usually begin with *puszcza.* English words 'wood' and 'forest' theoretically encode differences in size – the former being smaller, which is lost with the application of the plural. Woods can be as big as forests. Practically, their usage depends on individual choices. And one cannot be sure whether the selection of 'wood' invariably makes the property [SMALLER THAN FOREST] more salient. In both cases, the degree of specification is at its lowest.

Gołębie

It functions as an umbrella term for pigeons and doves, both of the same hierarchical rank in the English lexico-semantic field. Scientifically, the category groups four species occurring in Poland naturally: the wood pigeon, stock dove, collarded dove, turtle dove and the familiar but artificially introduced feral pigeon, as well as domesticated varieties. In Polish, only the feral pigeon retains the group nominal *gołąb* (the singular of *gołębie*) in its name: *gołąb skalny* or *miejski*, the rest being, respectively, *grzywacz, siniak, sierpówka, turkawka.* In English, two species are pigeons, three doves. As a rule the Polish schematic *gołębie* requires specification in translation. When embracing the whole family, one must say pigeons and doves, whereas when thinking about a flock of collarded doves, only 'doves' makes sense. City squares across Europe tend to support substantial populations of pigeons rather than doves. Each time a different element within the conceptual base must be profiled. Colloquially, all the species mentioned above merge in Polish into one basic level type: *gołąb.* While to translate a simple sentence such as 'Przyleciały gołębie i zaczęły jeść pszenicę naszykowaną dla kur' – '(*Gołębie*) have flown in and started eating wheat prepared for chickens' one can write without much risk 'pigeons' or 'pigeons and doves' or 'doves' (if collarded doves feeding together with the chickens happens to be a more familiar sight in the described place), in more specialised contexts all clues must be explored for correct profiling.

Grąd

Sawicki derives the name of the forest type called *grąd* from the old word for a rise among wetlands, stating that this type often accompanies riparian forests but occupies more elevated terrain (2003b: 20-21). In English, such forests do not have any special name. The most typical variety is labelled oak-hornbeam forest. Yet, the translator of Sawicki (2003b)[41] uses a quasi-term 'dry-ground forest' to denote *grąd*. When the author proceeds to describe different kinds of *grąd* and mentions *grąd niski,* rendered as low dry-ground forest the name does not sound too convenient because it is supposed to stand for something not dry at all. As documented in my database, dry-ground forest and wet-ground forest enjoy bewildering popularity with translators. Actually, the Polish *grąd* can be translated into English in at least about a dozen ways since each type requires adding the relevant tree or characteristic herb layer species name to the basic 'oak-hornbeam forest'. The highly schematic *grąd* must be replaced with a specific instantiation in English. Although he expression's profiles in both languages seem the same, in English one cannot avoid enhancing the salience of specific elements of the designatum (oak-hornbeam-linden forest, oak-hornbeam-linden-ash forest or even oak-pine forest, etc.). In Polish, these elements are only implied, and can become explicit if necessary. In most vague contexts, 'oak-hornbeam forest' seems the safest option. The Polish term, though taught at school and present in guidebooks, brochures and albums on nature, does not readily activate appropriate images in people's minds. Elderly country dwellers may go for 'a not necessarily wooded mineral swell among wetlands' as an immediate association. Paradoxically, a handy short lexical item, borrowed by science from folk terminology, represents the specialist's perspective to a greater extent than the layman's.

Kępa/kępka

It is an immensely polysemous word, expressed in English through a variety of lexemes. Translators from Polish keep forgetting about this, which gives rise to shameless mistranslations. In Polish, profiling is done by means of collocation or contextual clues, whereas in English by an employment of the right word. The most well-known collocations are *kępa drzew/krzaków/roślin/trawy/kwiatów* – a clump of trees/bushes, a patch of

---

[41] The translation of this book, as of many others cited in my book, is published in the same edition on the pages adjoining the corresponding Polish text (see Notes to the Reader).

plants/flowers, a tuft/tussock of grass/ a nest of (a specific species of plant). In these cases the diminutive *kępka* can be used to signal the small size of a given clump or tussock. For Poles living in the neighbourhood of the Vistula river, which despite its size escaped extreme regulation, *kępa* evokes images of a small island on the river but near the mainland, and of low flat land near the river. They have not been blurred, which often happens, by widespread adoption in the names of villages (Kępa Nadbrzeska, Kępa Wólczyńska, Kępa Gostecka, etc.). On the contrary, since these villages lie on 'a flat land near the river' their names enhance the meaning of *kępa* as a common noun. The equivalent, chiefly used in British English, is 'holm'. For the second meaning Americans use 'bottoms'. The trouble is that the relevant images do not spring automatically to the mind of a native speaker of English to the extent they do in the Polish mind. Sometimes it could be more salient to resort to less succinct verbalisations such as 'isle/islet on the river' for the first sense and 'floodland by the river' for the second. From a more specialist perspective *kępa* (*mineralna*) is a mineral island/swell amidst peat-underlain wetland, with a folk name *grąd/grądzik*, today largely incomprehensible to urbanised society. The degree of expert perspective still increases in the context of phytosociological studies of mires. Some of them have a hummock-hollow structure (*struktura kępkowo-dolinkowa*), in which *kępka* corresponds to 'hummock' – a little rise above the surrounding depression mainly formed of mosses and plants taking advantage of slightly drier conditions than on the lower mat of vegetation immersed in water. The English lexico-semantic field outsizes the Polish two-item category by far, concurrently permitting the development of a rich conceptual structure in each system.

Komary

Durrell writes: 'The first shoal of gnats, mosquitoes and lady-birds' (1975b: 201). In the Polish edition, gnats are rendered as *komary*, mosquitoes as *moskity* (Durrell 1999: 213). In English, gnats and mosquitoes are names borne by members of the *Culicidae* (*komarowate*) family of the suborder *Nematocera* (*długoczułkie*) of the order *Diptera (muchówki). Moskity (*subfamily *Phlebotominae* – sand flies*)* belong to a different family within the same suborder – *Psychodidae (ćmiankowate –* moth/drain flies*).* They suck blood and transmit dangerous diseases, like mosquitoes. In sum, in Polish, mosquitoes and gnats are both *komary*, *moskity* being sand flies. The name 'gnat' is more British, Americans prefer 'mosquito' but it seems that the British are increasingly like

Americans in their preferences concerning this name. Durrell, as a naturalist, makes fine distinctions in his descriptions and it is always worth recommending the mirroring of it in translation. However, in the case of gnats and mosquitoes, the splitting of the Polish umbrella term *komary* must not be sought at any cost. Such undue determination breeds inaccuracies of the sort exemplified above. Incidentally, in English everyday usage hardly anyone bitten by a gnat or a mosquito bothers to go beyond a schema, picking the names at random. Neither instantiation is profiled, to the effect that each of the two labels stands for the whole group – exactly like the Polish *komary*.

Krwawnik

Having two quite distant referents it may be thought of as a verbalised category correlating with two conceptual categories of clear-cut boundaries. One is a plant – the yarrow, the other a mineral – the cornelian. They can hardly be confused. And yet, dealing with translation, we prefer to approach the matter with less confidence. Let us consider the following: in *The magus* one of the female characters, Lily, during a performance staged by Conchis had nipples 'the colour of cornelians, as if they alone in all that honeyed skin, had been, or could be bitten and bruised' (Fowles 1977: 528). In Polish, we have: 'Sutki koloru krwawnika, na tle miodowozłocistej skóry, wyglądały jak pogryzione, poranione' (Fowles 1982: vol. II, 168). On the lexical level, the translation is correct but I recommend using the other name given to the mineral – *kornalina* to avoid wrong profiling. It is difficult to estimate what percentage of people know *krwawnik* as a herb and what as a mineral, but the number of those familiar with both is certainly small. I suspect, though, that the herb, once in folk medicine applied against haemorrhages would get more votes. The snag is that this common plant has white, sometimes light rosy flowers and in no way resembles female nipples, especially bruised ones. The readers who retrieve from the cognitive base such an image are indeed perplexed. What they are supposed to imagine is the blood red cornelian. If in the Polish text *kornalina* appeared, they would either assume that it is something blood red and continued reading undisturbed or would check this foreign-sounding lexical item in a dictionary. The moral learnt in this way is that, although the word *krwawnik* is suggestive of blood (*krew*), it can evoke images that diverge too much from the desired one. But even where the likelihood of confusion is lower, as in 'She wears now a small throat necklace of cornelians, the colour of dried blood' (Fowles 1985: 47) – '... mały naszyjnik z krwawników' (Fowles 1997: 51), *kornalina* fits better as a name for a precious stone, most such names having a foreign flavour. Remarkably, neither 'cornelian' nor 'yarrow' relate to blood in their verbal representations, which further justifies the choice recommended above. For the English reader the salient element is the colour. The profile of the word cornelian does not have a figure status. The translator should ensure that, in Polish, the two profiles of the word *krwawnik* do not gain salience over the colour in the scene.

Łęg

Writing about the Vistula river Szymon Kobyliński states that 'W ścinaniu fali powodziowej ... biorą udział łęgi. Łęgi to tereny w dolinach rzek, użyźniane przez zalew, świetnie nadające się na wysokodajne łąki i pastwiska' (1984: 54). – 'Marshy meadows can help to check the wave of flood waters. These marshy meadows in the river valleys made fertile by alluvial deposits and moisture are excellent as highly productive fodder-producing meadow or pasture' (ibid.: 54). Nowhere does he mention that *łęg* is a type of forest, accompanying rivers, brooks or underground watercourses, too. It figures in the terminology of forestry as well as grassland management studies and in the everyday lexicon, where it is understood as a periodically flooded or otherwise wet area where grasses and herbs are interspersed with shrubs and an odd willow or poplar tree, perhaps also a damp wood with an impenetrable undergrowth. Locally, *łęg* continues to signify places which used to fit this definition but, drained, are now covered with grasses and bushes tolerant of drier conditions. In English no such three-branched label exists. Thus, there are 'marshy meadows', or more precisely, 'floodplain meadows' and several names, sometimes descriptive, to denote woodland[42]. Largely expert jargon, their level of conventionalisation ranges from low to very low. The closest to folk vocabulary they get is 'riverside forest'. In one name, fen carr, they merge with what the Polish language carefully separates as *ols* (alder swamp). Notably, apart from the short handy label, Polish also has various descriptive names, for 'floodplain meadows' – *łąki zalewowe*, for 'riparian forest' – *podmokły las nadrzeczny* and so forth. To recapitulate, the maximal scope of the conceptual category of *łęg* corresponds to the scope of at least two distinct concepts in English. These concepts yield multiple verbalisations – boundaries of most categories they delineate overlap, in a few cases completely. In both languages the number of concepts and words which denote them diverge.

[42] See 2.2 pages 34, 36.

Matecznik/ ostęp

These two concepts totally overlap in one meaning: 'an almost inaccessible place in an ancient forest hiding lairs of wild animals'. The second of them additionally means 'a unit of forest division for the purposes of logging'[43], which a professional source *Załącznik do Zarządzenia nr 43 Dyrektora Generalnego Lasów Państwowych z 18. 04. 2003 r.* 1.1.4. [The Annex to the Regulation no. 43 of the Director General of State Forests of 18 April 2003 1.1.4.] specifies as follows: *obręb leśny* should be divided into *ostępy* (plural of *ostęp*) and that these may consist of e.g. 1-2 *oddziały* (pl. of *oddział*). This is worth mentioning because the two uses are often confused by translators. The English language does not have a one-word label for the first concept. As the database collected for this book shows, the highly conventionalised, compact way of verbalising at the same time [INACCESSIBILITY], [IN BIG FOREST], [WITH WILD ANIMAL LAIRS] in Polish is confronted with numerous circumlocutions, which rest on the immediate context. 'Lairs' appears in them probably most frequently. Among other renditions, the following crop up: 'bagienne ostępy' – remote marshlands, 'w ostępach leśnych' – in wild forested areas (Kłosowscy 2005: 19, 21), 'ich największym matecznikiem' – their [the European bison's] largest habitat (Sobociński 2004: 20), 'prawdziwie puszczańskim ostępom' – wooded and truly wild regions, 'puszczańskie ostępy na równi z lasami zagospodarowanymi' – wild groves and cultivated woodlands (Kłosowscy 1999: 98). Not all of them are recommendable as they stand but they unfold what may turn into a long list of possibilities. 'Prawdziwie puszczańskim ostępom' after correction could read as 'truly wild and inaccessible corners of the forest', 'puszczańskie ostępy na równi z lasami zagospodarowanymi' as 'pristine woodland and managed forest'. Some translators seem too fond of the word 'backwoods' to denote *matecznik* or *ostęp*[44]. It was first used by European settlers in America, who kept to the eastern coast, calling the wild and inhospitable interior in this way. A quick look at dictionary definitions should dissuade those translators from the word's blind application:

---

[43] As defined by http://www.sjp.pwn.pl, translation mine.

[44] See, for instance, Kossak (2001).

Webster's:
1. the wooded or outlying and only partly cleared areas on the frontier or in the backcountry
2. a rural area that is provincial or backward in culture or remote from the main centres of civilisation.

(www.answers.com/topic/**backwoods** - 44k):
1. heavily wooded, uncultivated, thinly settled areas.
2. an area that is far from population centers or that is held to be culturally backward.

(www.yourdictionary.com/**backwoods** - 17k):
heavily wooded areas far from centres of population; any remote, thinly populated place.

All of them miss the vital ingredient of meaning [WITH WILD ANIMAL LAIRS] and add others [THINLY SETTLED] and [CULTURALLY BACKWARD]. Cognitively, we can say that, in Polish, *mateczn*ik and *ostęp* automatically retrieve from the base three properties of their category. They are schemas, whose instantiations can vary significantly. One conceptualiser envisages an extensive alder swamp, another montane beech and fir forest studded with rocky outcrops or lowland coniferous stands with tangled and thick understorey.

The technical forestry usage of *ostęp* entails the change of perspective and resultant imagery, brushing aside all the three components of meaning. The operation of transferring *mateczn*ik and *ostęp* into English requires the decision which element of meaning ought to be profiled: the size, inaccessibility or primeval character of woodland, or perhaps, its capacity to hold healthy populations of large, wild herbivors and predators. Returning to the role *ostęp* plays in forest administration, it must be stressed that the divisions found in American and Canadian forests (the UK seems totally incomparable) are not uniform and in the first two countries usually stop at higher levels, with the smallest management units the size of the Polish *leśnictwo* or *obręb*. Thus, there are blocks split into compartments or management units, units composed of sub-units, ranger districts without subdivision, etc. If conclusions drawn on the basis of official documents issued by international organisations, written in good English but dealing with Russian forestry, can be valid, then the following correspondences hold true: *leśnictwo/ obręb* – forest district, *ostęp* – technical section, *oddział* – ranger compartment; alternatively, *obręb* may be a technical section, while both *ostęp* and *oddział* – a ranger compartment[45]. As long as *ostęp* – a forest administrative unit is not mistaken for wild animal lairs or the most inaccessible natural sites, the translator does not have to worry too much over the choice between units or sections. Notwithstanding different conventions and perspectives, the context should provide enough clues to convey the right message at an acceptable level of specificity.

Miedza

It is a ridge of land left unploughed as a dividing line between fields belonging to different owners. It derives from *między* = between. For Polish peasants, living from hand to mouth, it was almost sacred. Their tiny plots of land had to be clearly separated from the fields of their neighbours. Disputes over ploughing such borderlines were commonplace. The word has been fossilised in idioms such as 'spać jak zając pod miedzą' – 'to sleep like a hare by the balk'. The lexico-semantic field consists of practically this one item in Polish, which attests to its high degree of conventionalisation. In English, there exists one word equivalent, which theoretically indicates a similar level of conventionalisation but 'balk' can be a ridge of unploughed land left through carelessness not purposefully as a boundary. In Polish, no such extension is possible. Besides, 'balk' has several other meanings and does not seem to occur in relevant contemporary literature. Instead, two item expressions or descriptive labels appear: field boundaries, field edges, the edges of the fields, field margins, boundary features. The prototypical *miedza* is a long, straight and narrow – ca. 0.5 metre strip of herb vegetation with an occasional bush or tree (but it is not essential), located at intervals of even as little as twenty metres. Traditional Polish cropland fields are long and thin. The prototypical field boundary is a hedgerow, dry-stone wall, fence or a wider 1-4 metre-wide strip of grass contouring expansive farmland. The last possibility has entered the British landscape recently, promoted by agri-environmental schemes aimed at easing the strain intensive agriculture puts on the environment. *Miedzas* in Poland are culturally entrenched, so agri-environmental schemes just offer to let them survive the intensification of farming methods.

In cognitive terms, we can say that *miedza* projects a well-conventionalised image, much less schematic than any of the English

[45] See http://www.iiasa.ac.at/Research/FOR/forest_cdrom/english/for_man_en.html - 19k - accessed 29.12.08.

equivalents, rendering salient features such as [BETWEEN TWO FIELDS], [LONG], [NARROW], [LOW], [VEGETATED], [REGULAR IN OCCURRENCE ON FARMLAND] and has a neutral perspective preserving the same sense in specialist and everyday language. English equivalents are at a high level of schematicity and have a lot of possible instantiations. In various contexts radically different features can receive salience, 'boundary features' having the broadest maximal scope. Translations from English may not always be into the word *miedza*. In the context of proposals for more eco-friendly agricultural practice *obrzeża pól* or *granice pól* may be more pertinent. But where in the Polish text *miedza* is encountered, the designatum is unambivalent. How should it then be rendered in 'chodzili miedzą między saradelą a burakami' (Iwaszkiewicz 1969: 252)? The translator wrote 'walked along the boundary strip between bird's-foot and beets' (Iwaszkiewicz 2002: 178) – a good choice. 'Balk' would also be suitable. All the other expressions listed above miss a vital part of the meaning – we are not talking about the edge of the whole field or at any of its sides but about a strip of unploughed land between two crops. Whether they belong to one owner or not is unspecified, though prototypically they do. We assume the strip is thin. No such assumption is inherent in: field boundaries, field edges, the edges of the fields, field margins, let alone – boundary features. All in all, despite differences in the degree of conventionalisation at the three remaining levels, imagery can be adjusted in Polish and English through the manipulation of lexical items. With other usual components of meaning latent in the Polish sentence, 'boundary strip' sufficiently embodies those brought out – 'thinness', 'betweenness' and 'being unploughed'.

Ostoja

In its colloquial usage, its English equivalents – (wildlife) sanctuary, (wildlife) refuge correlate closely in respect of conventionalisation, schematicity and perspective. In both languages the words primarily refer to human affairs, evoking such states as a sense of security and/or stability. Salience, however, differs in one crucial aspect. In English, the feature 'place' is central, in Polish *ostoja* could be a place as well as a person. Foregrounded are the feelings, not structures or areas. The problem arises when this strongly rooted folk gestalt perception is exposed to technical nature conservation terminology. The term *ostoja* was borrowed in the 1990s by Polish conservationists to name areas identified as particulary favourable for birds within the Natura 2000 network. Prior to the implementation of this EU programme the term functioned in the State Forests (*ostoja zwierzyny* = wildlife refuge) as an informal designation of sites where wild animals, both game and those protected by law, supposedly or really mated, reared their young or just habitually hid and foraged during the day. Such sites, having no formal borders and no clear criteria of selection (cf. e.g. Kossak 2001: 483), have been perceived less as specific places than as a general idea. Propagated by foresters aiming to discourage disturbance, and at schools, it suffuses Polish minds as a firmly established but rather vague concept. In the new conservationist meaning it denotes far larger fragments of much more diverse types of landscape officially designated in order to ensure the survival of especially valuable species by maintainig their habitats, tightly knit to the whole ecosystem. It corresponds, *inter alia*, to the English IBA (Important Bird Area), an official term used by BirdLife International, the organisation which prepares inventories of such areas in different countries. In Europe IBAs became the basis for the designation of Special Protection Areas within the Natura 2000 programme stemming from the 1979 Birds Directive. Interestingly, in Chylarecki and Sawicki, in the Polish text, 'tereny ważne dla ochrony ptaków' (2003c: 10-11), literally 'important bird protection areas' is used instead of 'ostoje ptaków' ('bird refuges') when no reference is made to the areas officially designated. In English, one can observe the reverse situation. 'Bird refuges' are used descriptively, IBAs in the context of officially set up *ostoje* (plural for *ostoja*). The depth of cultural entrentchment of the concept of *ostoja* as an idea rather than a place, strengthened at the lexical level by the fact that it collocates with the words 'obszar, teren' can be illustrated by a hilarious mistake made in Sawicki. 'Na terenie ostoi' was translated as 'within the area of Important Bird Area' (2003b: 74-75) – an automatic transposition without due regard to the resulting tautology.

Podszyt, runo, poszycie

In specialist terminology Polish *podszyt* (*podszycie*) has one conceptual equivalent – 'shrub/bush layer' (of forest), *runo* correlates with two: 'ground layer' (mosses and lichens) and field (herb) layer[46]. The Polish words also function as non-specialist terms. The same lexico-semantic field of layers of vegetation in woodland contains one more item: *poszycie*, which does not figure in expert texts. In folk terminology it equals *runo* in

[46] Compare Szymański (2000: 169), Dunn (1990: 13) and the *Book of the British countryside* (1981: 524).

meaning[47]. While concepts behind Polish lexemes have relatively clear boundaries, lexemes themselves evoke rather vague images in laymen. English verbalisations of the same categories are much less schematic than Polish ones and there is no doubt what they profile. However, the corresponding English lexico-semantic field stretches much further and for the shrub layer – understorey, underwood and brushwood can be substituted. Although Dunn considers understorey identical with 'shrub layer' (1990: 13), in common use it encompasses the three layers: ground, herb and shrub. In addition, a category of undergrowth exists, defined as a low growth on the floor of a forest including seedlings and saplings, shrubs, and herbs[48]. All of them can be met in expert works such as *The management of semi-natural woodlands* (2003) or the *Interpretation manual of European Union habitats* (2007)[49]. The multiplication of verbalised categories certainly makes the work of the translator harder, but by foregrounding slightly different aspects of the same fragment of reality, they take the credit for stylistic enrichment. Highly conventional *poszycie, runo* and *podszyt*, understorey, undergrowth, underwood and brushwood have lost the sharpness of the two-element and thus less conventionalised lexical units. Terminological fields of specialist Polish and English have one more item normally absent from the everyday lexicon: *dno lasu* – forest floor, which in both languages cover the ground and herb layers. Since 'forest floor' is self-explanatory, in certain non-specialist contexts it can serve as an acceptable rendition of *runo*. Mismatched registers, so by implication – perspectives, pale into insignificance at the opportunity to be brief and precise. As the database I have compiled indicates, translators from Polish seem to favour the underspecified 'undergrowth' to stand for *podszyt* and, more curiously, for *runo/poszycie*. In narratives, such a change at the level of salience and schematicity rarely matters but in guidebooks or albums it may not be safe and desirable.

Rogi

Properly speaking, it stands for horns, their essential property being a bony core and a keritanised sheath. *Rogi* is an attribute of cattle and their wild relatives. Horns are permanent and unbranched. Antlers are a pair of bony, branched structures that protrude from the frontals of the skull of

---

47 Confirmed by the PWN online dictionary at http://sjp.pwn.pl (accessed 20. 12. 08).
48 http://mw1.merriam-webster.com/dictionary/**undergrowth** (accessed 19. 12. 08).
49 Available at http://ec.europa.eu/environment/nature/legislation/**habitatsdirective**/index_en.htm - 32k

animals of the *Cervidae* family and are shed annually[50]. The latter is *poroże* in Polish. It so happens that in English the two terms are commonly used interchangeably as if in defiance of scientific categorisation sensitive to differences but in agreement with the ordinary human's tendency to seek similarity in diversity. In Polish, *poroże* represents the specialist usage by hunters, foresters, breeders or biologists, hardly ever heard from the layman's mouth, for whom *rogi* fill the relevant lexico-semantic field singly. To further reinforce the specialist status of *poroże* the bony material when used in the tools of our early ancestors or in sculpting, is invariably called *róg* (only singular). In translation, an informed choice tailored to a particular context must be made whether 'wykonane z rogu' should become 'made of horn' or 'made of antlers'. In other words, profiling is a necessity, where in Polish the unspecific maximal scope is relied on.

Sitowie

A general Polish dictionary definition is 'a plant with flat and rough leaves, which grows on water edges and in wet meadows, forming dense stands'[51]. The botanist would point to one species *Schoenoplectus lacustris – sitowie jeziorne*, currently *oczeret jeziorny*, while the layman extends the name over any aquatic or wetland plant that in his/her opinion looks approximately as described above. To the botanist's dismay (and mine as well) even the reed can fall prey to such extensions. It was done by the translator of *Daniel Martin*: '[Mark] climbs down into the reeds – '[Mark] schodzi w sitowie' (Fowles 1978: 33). If translators consulted one of the newer expert sources, they would be disappointed to learn that no plant of this name exists – former names are rarely included. And yet, it regularly occurs in older texts, especially narratives. In English, the matter becomes even more complicated. In the past, the name bulrush was often attributed to *Schoenoplectus lacustris.* According to Blamey, Fitter, and Fitter (2003: 342) it was later officially given to *Typha latifolia*, alternatively called reedmace and known in America as the cattail – *pałka wodna. Schoenoplectus lacustris* is now the common club-rush. In most literary descriptions *sitowie* has a highly schematic, folk meaning. In nature albums or guidebooks, as they often try to compromise specialist and non-expert perspectives, both the schema and its instantiation can appear. As a

---

[50] See http://animaldiversity.ummz.umich.edu/site/topics/mammal_anatomy/horns _and_**antlers**.html - 23k (accessed 28 Dec. 2008).

[51] As defined by http://sjp.pwn.pl, translation mine.

schema it can be rendered as 'rushes'. An instantiation in the translation from English might force the translator to resort to guesswork. For example, in *The French lieutenant's woman*, the following sketchy description is stripped of any clues: 'A *bas relief*, small sculptures, an urn with bulrushes' (Fowles 1996: 425). Whether the urn contains *Typha latifolia (pałka wodna)* or *Schoenoplectus lacustris (oczeret jeziorny)* remains a mystery. Since the two plants look distinct, the visualisation of the decorative elements in the room lends itself to different interpretations. In this case, the translator opted for the latter and its former name: 'Płaskorzeźba, małe rzeźby, urna z *sitowiem*' (Fowles 2003: 404). Now, if we assumed that the Polish version was the original, the contemporary conscientious translator would not come up with 'bulrushes' in English but 'common club-rushes', perhaps abbreviated to 'club-rushes'. Evidently, to accommodate all the category members mentioned, terminological and lexico-semantic fields in both languages have to take into consideration diachronic as well as synchronic data.

Szuwary

For non-specialists they are stands of tall plants other than trees or bushes, fringing lakes, swamps and the like. Experts differentiate numerous types on the basis of dominant species. The most common is *szuwar trzcinowy* (reed swamp), with the bulrush swamp (*szuwar szerokopałkowy*), common club-rush swamp (*szuwar oczeretowy*) and reed sweet-grass swamp (*szuwar mannowy*) following suit. In practice, *szuwary* can be either reeds or rushes or reeds and rushes. Calling them, logically, swamps, would foster ambivalence because swamps can denote an entire natural habitat not just its vegetation and can refer to alder forests, which by no means resemble *szuwary*. It can also signify tall sedge swamp (*szuwar wysokoturzycowy*), which overlaps with another ambiguous term *turzycowisko*. The ambiguity stems from the fact that some specialists, apparently refuse to approve of the narrow understanding: tall sedge swamp, embracing small sedge beds, too. Translating *szuwary* as sedges or sedge beds is inappropriate for the same reasons as in the case of 'swamps'. The category which comes closest to *szuwary* is *trzcinowisko* – reed bed or, for short, reeds. Obviously narrower in scope, in impressionistic, as a rule imprecise, descriptions it suffices and rarely errs on the side of underspecification. Whereas in Polish there exists an umbrella term of a high degree of schematicity, English lacks it. Each time the translator has to decide whether *szuwary* in question are reeds, rushes or a mixture of both or tall sedge swamps. If the vantage point shifts to that of the specialist, which happens in narratives too, the number of possibilities increases dramatically.

Tobołki

This is a curious category since it corresponds to four very different concepts and yields four incongruous lexico-semantic fields. The first, familiar to all Poles, is 'large, cumbersome packages in the form of things wrapped up in a piece of cloth'. The second brings to mind a group of plants – penny-cresses, with field penny-cress (*tobołki polne*) being the most common both in Poland and in the UK. The third concept is confined to north-eastern Poland but has recently been popularised with the advent of ecological education combined with a rise in visits to national parks. The name signifies small dystrophic lakes fringed with transition mires, such as those scattered around Puszcza Augustowska, mainly in the Wigierski National Park. The last and least known referent is the phylum of algae – *Dinophyceae*, in Polish also called *dinofity*, in English dinoflagellates. The maximal conceptual scope can impress us. Nevertheless, the likelihood of confusion is low, although *tobołki* – dinoflagellates can potentially be pulled out of *tobołki* – the lakes. The term cannot function as a schema, each usage bringing particular instantiations. Only the first meaning reflects the layman's perspective, the remaining three used either regionally or by specialists. English words correlate with the Polish equivalents almost exactly in respect of perspective. The main difference, facilitating the translator's work, is that none of the lexical items relates to one another in any way.

Turzycowisko

As already mentioned under the *szuwary* entry, even expert usage lacks consistency. Most restrictively, *turzycowisko* denotes tall sedge swamp. But some specialists extend it over sedge beds/communities/fens. The choice depends on the optics: 'beds' is most neutral, 'communities' more botanist-oriented, 'fens' favoured by geo-botanists and peat specialists. In fact, boundaries between tall sedge swamps and sedge fens tend to be fuzzy, the transition zone stretching and shrinking according to local conditions. Thus, small sedge fens can contain an admixture of tall sedges and vice versa (see Chapter 2.3). In contemporary 'urbanised' folk terminology *turzycowisko* seems barely recognised. If at all, it is used in its broader sense. In brief, in Polish and in English the term is generally viewed from the detached, specialist perspective. Although the lexical item representing the conceptual category sounds well conventionalised and by no means alien, it does not go beyond evoking a very vague schema in most people's minds. By contrast, specialists immediately imagine a range of instantiations. On their classification the survival of a given natural site hangs. If they choose tall sedge swamp as the profile of *turzycowisko*, the site will not be protected within the Natura 2000 system, while if the site is declared 'sedge fen', it gains conservation status.

Uroczysko

It has three folk meanings, all of them forgotten by town dwellers today and younger generations of country people and replaced with the imagery that directly relates to the contemporary understanding of its root word. The term survives in topographical names on maps, old books and publications on traditions of people inhabiting woodlands and marshlands of Poland on its former and current territory. Hence the older meanings circulate, repeatedly revived in the minds of nature or history enthusiasts. They are as follows:

1. inaccessible terrain, usually swampy and wooded; also: secluded place.
2. a fragment of larger area with natural frontiers, distinguished through a topographic name.
3. in ancient Slavs' terminology: a site in the heart of primeval forest, where they worshipped their gods, held meetings or which they considered a seat of evil spirits[52].

The people of Polesie, the region once partly Polish, at present stretching on both sides of the border between Belorus and Ukraine have a very concrete image behind the concept, a specific instantiation of the schema projected by the three quoted dictionary meanings. This is how Józef Mackiewicz describes it in *Bunt rojstów* (cited in Rąkowski 2001: 285, translation mine):

> Ostrowy, albo, jak je nazywają chłopi, uroczyszcza, są to małe i większe pagórki, oazy wśród błot. Raje, pełne rozświergotanego ptactwa wśród mieszanych zagajników, drzew iglastych, dębów, głogu, jarzębiny, leszczyny.

[52] Explanations from http://sjp.pwn.pl, accessed 29.12.2008, translation mine.

> *Ostrowy*, or as they are called by local peasants, *uroczyszcza*, are smaller and larger hillocks, oases among the swamps. They are like a paradise, full of singing birds which sit in young-growth deciduous and coniferous stands, needleleaf trees, oaks, hawthorns, rowans and hazel scrub.

The word *ostrowy* in Old Polish meant 'river island' or 'small lake island' (Maćkiewicz 2004: 202-203).In contemporary Polish it has survived only in placenames but it remains part of the present-day lexicon of Eastern Slavs. *Uroczyszcza* (plural) is an 'easternised' version of *uroczyska.* Wiśniewska explains that *uroczyszcze* is of Ukrainian origin and means 'wild, mysterious place' (2001: 306). Nowadays people tend to think that *uroczysko* denotes an exceptionally beautiful natural spot (*uroczy* = charming, delightful, enchanting, beautiful). Since on average post-industrial societies do not judge as beautiful inaccessible swamps or impenetrable thickets of forest undergrowth and hardly ever reach truly secluded sites, such a spot can lie two miles or so from the nearest car park and can take the shape of a birch grove with nice soft grass to sit on, or a sunny hillock carpeted with flowers, guarded by imposing lindens and oaks, and sloping towards an open valley. Meanwhile, *uroczysko* derives from the old folk word *urok:* magic power that can do harm to somebody or something. Evil forces go hand in hand with charm – beauty can prove deceptive. Both understandings meet in the original concept of *uroczysko* – awesome views and pleasing sounds of nature border on the mysterious, the dangerous and the treacherous.

Rysiewicz reminds us of yet another, more technical meaning of the word: 'Puszcza dzieliła się wtedy na ostępy i uroczyska, bory i półborki ...' (2000: 16) – 'At that time the forest was divided into various units called *ostęp, uroczysko, bór* and *półborek*' (translation mine). The original translation[53] contains too many errors to be discussed in this section. *Uroczysko* becomes 'range' there – employed as an equivalent in the majority of cases we have examined, despite its rather limited application in forest administration in the English-speaking world. As a forest management unit it probably functions only in India, normally referring to a series of mountain tops, extensive unenclosed areas where livestock grazes or simply, places where one can roam freely. The word 'range' often accompanies a proper name in translations from Polish (cf. e.g. Kossak 2001) since forest management units and outstanding landscape features often obtained names for better communication. Instead of

[53] 'A primeval forest was divided into backwoods, ranges, woods and half-woods ...' (Rysiewicz 2000: 16). The fact that woods are not *bory* is of secondary importance here.

writing, 'Bukraba's ranges' – 'areas owned by Bukraba' could appear. 'The Grzybowiec range' could be replaced with 'the site called Grzybowiec', 'the Kniehinine Błoto range', in turn, with 'the marshes/mires called Kniehinine Błoto', etc. As demonstrated, the repertoire of ways in which *uroczysko* can be transferred into an English text is as rich as an inventory of instantiations. Where the Polish language chooses to remain vague, English has to profile a presumably most pertinent property. The translator is obliged to do his/her best to find out whether putting 'an ancient sacred site called ...' or 'the in-swamp hillock called …' matches the reality. What if there is no palpable reality behind the word, as in the legend under a map? One is confronted with non-specific, non-referential *Ur. – Uroczysko.* The translator can only copy the abbreviation, attaching bracketed information – 'put before names of old forest units or landmarks'. To sum up, the Polish word *uroczysko* is culturally loaded, therefore requires a profound effort on the part of the translator. The perspective of its usage ranges from professional to lay, from regional to standard, from historical to modern. It varies in the degree of conventionality and schematicity, different members of the category encoded in this lexeme gain relative salience sometimes even in one and the same context.

Ważki

This name covers two suborders of the order of insects *Odonata*: *ważki różnoskrzydłe* (*Anisoptera*) and *ważki równoskrzydłe* (*Zygoptera*). In English the former are dragonflies, the latter damselflies. Put together, they stand for the order of *Odonata* (Gibbons 1999: 14-52). For the sake of precision and logic, whenever one generalises, it is necessary to write 'dragonflies and damselflies'. To the scientists' disappointment, logic and precision lag behind features valued most in folk terminology. Not just in casual talk but even in popular publications 'dragonflies' may encompass both suborders. As a result, peculiar statements such as the one below, exerpted from a children's book on nature, lose their peculiarity in the eyes of common people: 'The largest dragonflies of the present day are South American giant damselflies.' (Stidworthy 2001: 15). Translators from Polish, consciously or not, embrace the idea of messy simplification with glee (see for instance, Hryniewiecki and Okołów 2004: 119). In principle, we insist on the translator's responsibility for preserving scientific truth and rendering *ważki* as 'dragonflies and damselflies' when the degree of schematicity in the Polish text is high and using one or the other name if, by implication or ostensively, *ważki różnoskrzydłe* or *równoskrzydłe* are profiled. Certainly, there are situations where a more relaxed attitude can be adopted. Let us consider the following sentence from *Daniel Martin* (Fowles 1978: 24): 'Two shimmering blue demoiselle dragonflies with ink-stained wings flutter over the flowers, then drift away.', where 'blue demoiselle dragonflies' become 'niebieskie pannice ważki' in translation (Fowles 2000: 34). This translation of 'demoiselle dragonflies' is literal but acceptable since the author himself, usually very careful as far as scientific classification is concerned, uses his own, or perhaps, folk categorisation. Strictly speaking, this is the suborder of damselflies that includes the family of demoiselles. To be correct, we should say demoiselle damselflies then. There are only two demoiselle species in the UK, so writing just 'demoiselles' is a possible option too[54]. Both species of the family are called in Polish *świtezianka* (*świtezianka błyszcząca* – banded demoiselle, *świtezianka dziewica* – beautiful demoiselle). To sum up, the best translation of 'demoiselle dragonflies' would be *świtezianki.* Admittingly, Fowles himself does not prioritise scientific precision here, salience is granted to an impressionistic vision, the interplay of colours and movement of delicate wings.

Wodno-błotny

How deeply rooted in Polish minds is the image behind this adjective is illustrated by all too frequent translations of it as 'water and mud' and 'water-marsh'. Thus, we read about 'water and mud birds' or 'water-marsh environment'. The compound separates and simultaneously joins two pictures – that of water and of the muddy shore, an interface water/land. The English equivalent, wetland, omits the presence of water, focusing on the clayey or peaty flat land or silt-laden sands, which, it goes without saying, become wet and muddy when they come into contact with it. In other words, the Polish word conventionalises the automatic salience of two elements, the English verbalisation limits itself to one and by doing so, misleads Polish translators, as if compelling them to look for a more image-faithful rendition. If not inquisitive enough, such a search ends in awkwardness.

Wrzosowisko

The prototypical *wrzosowisko* is open landscape blanketed by heather. In Poland such habitats are never as vast as in the UK. They fill usually

[54] See http://www.dragonflysoc.org.uk

small gaps in tree stands, caused by fires, blow-downs and other catastrophes as well as exceptional soil infertility. Sometimes heather encroaches peatlands with disrupted water regimes, which while common in Britain, is quite scarce in Poland. The most noteworthy site of this kind is the Bielawskie Błota in East Pomerania. Military testing areas, where explosions, fires and movements of heavy equipment hamper the growth of trees, support most extensive heather communities on sandy soils. In English, there are basically two equivalents of *wrzosowisko*: heath/heathland and moor/moorland. Heaths occur in lowlands, moorland is upland overgrown with heather, gorse, grasses and lichens[55]. Both terms apply also to places in which grasslands dominate over heather scrub. Statistically, one such site contains more ling cover than most uninterrupted stands of the same undershrub in Poland anyway. When translating from Polish, one must remember that the term most likely to be used is 'heath'. Moorland, especially heather moorland practically does not exist in Poland, the nearest in appearance being *połoniny* in the Bieszczady Mountains, and yet the ling hardly occurs there. As shown in Chapter Four, this simple truth tends to be ignored in translation. The translators from English into Polish have it easier, 'moorland' and 'heath' can normally be rendered as *wrzosowisko* without running the risk of mistranslation. Hence, the following sentences from *A maggot* can safely have the same word in lieu of both printed in bold: 'In the distance could be seen a small town whose church tower they had made out from the **moorland** above' (Fowles 1985: 13) and: 'Why, this very first day we must pass those **heaths** of Bagshot and Comberley' (Fowles 1985: 142).

On the plane of verbalised categories, the correspondence of *wrzosowisko* and moorland/heath cannot be questioned. It is the concepts that cause trouble. For the Polish mind *wrzosowisko* is dry, small, tucked away in forest, with low-growth heather. For the British heath and, above all, moorland are usually quite sizeable, exposed to gales and heavy rains, inhospitable, easy to lose one's way in and devoid of trees for miles on end. Heather reaches in them impressive heights and is as impenetrable as the dwarf mountain pine scrub. Often underlain with peat, particularly on bare hilltops and slopes, they are soaked with water. They bring immediate, positive or negative, associations with red grouse hunting and rambling – in many private lands a forbidden fruit. They are unnatural but unique and cherished at the cost of native woodland, although wherever not managed for grouse – prone to transformation into pastureland or conifer plantation. Like Polish forests, in the past they hid fugitives,

[55] For a description see e.g. Brown (ed.) (1992: 130-132).

inspired folk imagination and impregnated poets' and writers' minds with mysterious and extraordinary ideas. In view of all this we can say that an Englishman always gives salience to many more components of the predicate's profile than are activated in the Polish mind. The schematic *wrzosowisko* must obligatorily be split into two – for the Polish people already instantiations, for the English still schemas. Whenever the context justifies it (in practice frequently indeed), less conventional renditions of *wrzosowisko* should be encouraged, for example, 'a small heather-clad glade', 'a patch of heather scrub', 'a site overgrown with heather' and so on. In turn, translating 'moors' as *bagniska* (huge/deep mires) may at times appear a better solution than the standard label – compare 'In a great wood we call Cleave Wood, that stretches to the moor and is more steep that ne'er a man may walk in many places, more cleave than combe' (Fowles 1985: 73) and 'W wielkim lesie, który my nazywamy Gęstym Borem. Rozciąga się aż po bagniska i rośnie czasem na takich stromiznach, że człowiek nie zawsze może tam przejść' (Fowles 1997: 74). The overall atmosphere of *A maggot* and the awareness of how boggy moors may happen to be allow us to believe that *bagniska* does a better job in the text than *wrzosowisko.* It is a matter of sensitivity to correct profiling and of the ability to reflect the best alternate construal of the scene.

Zarośla/Krzaki/Chaszcze

The first of the three sounds most scientific and literary, although occasionally it is used in everyday speech. The other two represent a lay perspective decisively, not infrequently having a deprecatory flavour. The most suitable English equivalents of *zarośla* are scrub and thicket(s), also brushwood – a thicket composed of shrubs and small trees. 'Brake' is an additional possibility. It can either refer to bracken, a huge invasive fern (*orlica*) or to any area of brushwood, thick undergrowth or thicket of one kind of plant. The last one should be approached with caution because no Polish category, conceptual or verbalised, profiles at the same time stands of ferns and woody plants. Most prototypical *zarośla* (a plural noun only) consist of various species of willow, while in dry sites – of blackthorn, hawthorn, brier and dogwood. Their prototypicality is a matter of degree to the point that 'willow' obligatorily collocates with *zarośla* not *krzaki,* whereas the other four take both. A thorny thicket, such as formed by blackthorns and briers can be called spinney but it has really fuzzy boundaries and could mean a small wood with undergrowth just as well. Besides, unlike scrub, thicket and brushwood it has not found its way into international English, remaining useful for literary and culture-bound texts. *Zarośla* occur in clumps or blanket substantial areas outside forests or on their margins. A woodland understorey is not as a rule referred to in this way, which makes 'brushwood' quite problematic as it does not discriminate between an underwood thicket or a thicket elsewhere. *Krzaki* (in specialist uses or for the sake of stylistic elegance – *krzewy*) are bushes or shrubs, *chaszcze* posing the greatest problem in translation. Essentially, they equal *krzaki* in meaning, perhaps with an emphasis on the bushes' density of growth but their choice in an utterance might suggest that the speaker disapproves of those thickets, discouraging penetration, concealing unknown threats, etc. The contrary can also be true – the speaker prides himself/herself on being able to trek through such thickets hiding nature's attractions. The application of *chaszcze* demands careful examination of contextual clues to give salience to the right interpretation of one's attitude by appropriate phrasing in English. The English language does not possess such an ambivalent one-item verbalised category.

Buntings

Deciphering the specific species on the basis of an enigmatic label in the following sentence does not take a naturalist much time: 'The reed beds attract a host of water fowl such as buntings ...' (the *Book of the British countryside* 1981: 381). The birds in question must be reed buntings and not any other buntings. Somewhere in the fields in Britain, however, one could meet during the breeding season a corn bunting, cirl bunting as well as reed bunting again – which inhabits sites other than reed beds with increasing regularity. In Polish, those birds do not bear a common 'surname' and we have to decide which species is meant: *potrzos* (reed bunting), *potrzeszcz* (corn bunting) or *cierlik* (cirl bunting). One should also know that during the breeding season the ortolan bunting, quite common in parts of Poland, does not occur in Britain, while outside it the number of bunting species rises to cover migrant or wintering birds – the snow and Lapland bunting (*śnieguła* and *poświerka*). The most curious thing about the name for the whole family of *Emberizidae* is that its most numerous representative is not called 'bunting' but 'yellowhammer'. In Polish, more logically, 'yellowhammer' (*trznadel)* lends its name to the whole family of *trznadlowate*. In English, the verbalised category of bunting almost entirely overlaps with the concept of it – all but one instantiation are lexically marked for inclusion in the group. In Polish, the family name is a schema, whose instantiations form separate verbalised categories. It is easier to say *potrzeszcz* than locate it within the suitable taxonomic group. In English, such an identification happens automatically. To a British person saying 'a bunting' comes more naturally than saying *trznadlowaty* comes to a Pole, who perceives it as belonging to specialist vocabulary. From a non-expert perspective 'a bunting', translated as 'jakiś trznadlowaty' sounds strange. To preserve the appropriate register something like 'chyba trznadel czy coś podobnego', literally, 'maybe a yellowhammer or something like that' should be written. Better still, if the context allows us to guess which of the bunting species is almost certainly meant.

Catchflies

Contrary to appearances, these are not insects but plants. A lot of members of *Caryophyllaceae* bear that name. In Anioł-Kwiatkowska (2003) two species of 'catchfly' are called *bniec*, one *firletka*, one *słonecznica*, one *smółka* and 24 *lepnica* (e.g. *dwudzielna* – forked, *dzwonkowata* – bell-shaped). The existence of such a capacious umbrella term in English ought to put translators from Polish on alert. Consider the following sentence and its existing translation: 'Południe już było gorące, świerszcze ćwierkały i pachniały czerwcowe smółki. Kwitły w rowie różowo.' (Iwaszkiewicz 1987: 367) – 'The afternoon was hot now, crickets were chirruping and there was a scent of June catchfly, blooming pink in the ditch.' (Iwaszkiewicz 2002: 133). *Smółka pospolita* is a very characteristic plant because a section of its stem is covered in a black sticky substance. This feature is immortalised in its Polish name (*smoła* = tar). The English name is sticky catchfly. Writing just 'catchfly' leaves too much space for interpretation. Most of the options can be rejected as they stand for species absent from Poland, where the quoted book is set in, or are rare. Having said that, these are not necessarily the same species that do not grow or are seldom found in Britain. Actually, the small-flowered catchfly (*lepnica francuska*) and the night-flowering catchfly (*bniec dwudzielny*) surpass the sticky catchfly in population and popularity with people in the UK. More importantly, while *smółka* is immediately recognisable by countrymen and nature lovers in Poland, a catchfly brings multiple images to the British mind, especially as most of them are various shades of pink. In fact, the comments made for buntings largely apply to catchflies: particular instantiations present themselves in the Polish mind much more readily than a schematic image of the whole family. In English the opposite seems true.

Chats

It is a collective name for eight British birds: the black redstart (*kopciuszek*), bluethroat (*podróżniczek*), nightingale (*słowik rdzawy*), redstart (*pleszka*), robin (*rudzik*), stonechat (*kląskawka*), wheatear (*białorzytka*) and whinchat (*pokląskwa*). All these species breed in Poland and some of their other relatives, too, but they have not got a collective name conventionalised in language. Sometimes they are grouped along different division lines: the black redstart and redstart as *pleszki*, bluethroat, nightingale and robin as *słowiki*, while stonechat and whinchat as *pokląskwy*. The shortest, though still longish translation of 'chats' into Polish is then 'pleszki, pokląskwy i słowiki'. In statements such as the one about the black redstart: 'this chat started to breed in Britain in the 1940s' the term can be omitted altogether without much loss to the text. Translators can be easily misled, though, by the presence of 'chat' in the names of two birds, in Polish collectively known as *pokląskwy*. Under no circumstances should they assume that this could be the name of the whole group. The category is defined by the sum of its members. Although fairly schematic in English, in Polish it requires a higher degree of specification, the enumeration of components arranged in subgroups.

Cherry

In Polish, we carefully distinguish between *czereśnia, wiśnia* and *czeremcha*. In English, 'cherry' has a rather flexible prototype stretching over the three categories. The *Book of the British countryside* (1981: 96-97) implies that the prototypical cherry is the wild cherry (*czereśnia ptasia*), whose fruits are sweet. It does not even give a mention to the dwarf cherry (*wiśnia pospolita*), which has tart fruit and does grow in the UK, devoting some attention to the bird cherry (*czeremcha zwyczajna)*. If one wants to make sure that they mean *wiśnia* – the cultivated one, they add before the word 'sour'. Sometimes *czereśnia* is referred to as 'sweet cherry', which is much easier to internalise for a Pole accustomed to separate names for the two fruit tree species. *Czeremcha* to the average Polish non-specialist is not automatically recognisable as related to the other two at all. A broad and a little fuzzy category of cherry compels the translator of an English text into Polish to carry out a difficult investigation before the actual transfer takes place. An unspecified cherry tree does not have any equivalent in Polish, in which such a schema has not been conventionalised. However, texts usually sooner or later reveal clues such as the taste of the fruit picked from this tree, its size, the texture of the leaves and so on. Then the language-imposed selection of instantiation becomes possible to a certain degree.

Copse/coppice/grove

The OALD, CED and Webster's treat 'coppice' as identical with 'copse'. OALD defines 'coppice' as 'a small area of woodland with thick undergrowth and trees'. According to the CED, it is a 'small group of trees growing very close together'. In Webster's it is 'a thicket, grove, or growth of small trees that are cut on a short rotation' and also 'a forest originating mainly from sprouts or root suckers as opposed to one derived from seed'. Blamey, Fitter, and Fitter (2003), however, separate 'coppice' and 'copse': 'Coppice: trees or shrubs, most often Ash or Hazel, cut nearly to the ground and growing again from the old stools. Not to be confused with a copse, which is a small patch of woodland'. Another related word – grove in the OALD is defined as follows: 'group of trees; small wood', in the CED: 'a group of trees that are close together. … *open fields and groves of trees*', and in Webster's: 'a smaller group of trees than a forest often without underwood and planted or growing naturally as if arranged by art'.

These three categories with fluid frontiers are often used to translate equally fluid but differently cut categories in Polish. While we can relatively confidently say that 'grove' correlates with *gaj/gaik*, coppice and copse have no conventionalised one-item lexical unit in Polish. Bilingual dictionaries tend to claim that the latter two are *zagajnik*. But in this word there are two inextricably bound defining properties: [SMALL] and [YOUNG] wood. The copse and coppice could be old, so the employment of *zagajnik* as their equivalent must be tailored to an individual case. The other, culturally-loaded meaning, is not even mentioned. A specialist dictionary (Czekierda 1995) rises to the occasion and we get to know that coppice is *lasek* (small wood), *zagajnik* (as already observed, it cannot be ruled out) and, importantly, *las odrostowy*, *zarośla ścinane okresowo*. The last two are a descriptive term and a brief description, respectively, denoting a traditional method of deciduous wood/scrub management in Britain, which although known in Poland, has never achieved comparable popularity and pre-thought systematicity. For the translator, it is paramount to profile the right image. A coppice in an English text stands a big chance of signifying 'a thicket, grove, or growth of small trees that are cut on a short rotation' – *zarośla ścinane okresowo* and also 'a forest originating mainly from sprouts or root suckers as opposed to one derived from seed' – *las odrostowy*. Writing simply –

lasek would detract from the salience of the text as a promoter of cultural realia. Translation studies call such an operation – domestication. Sometimes perfectly justified in narratives, it cannot be tolerated in publications such as guide books or albums on native nature, by definition aiming to present things and phenomena potentially perceived by receptors as curious, novel and foreign.

Cultivars and breeds

Generally, they squeeze the conscientious translator dry, forcing him/her to find out whether they are country-specific or have been successfully grown or bred in the target language culture. Establishing the names of species is easy in comparison with its human-made varieties. Certainly, one can risk leaving the source language name unaltered in translation, when suspicions are strong that the cultivar in question has not made a career abroad gaining new names. However, rendering Cox apples as 'jabłka Cox' instead of *koksa,* familiar to nearly all housewives, would cause problems with the identification of the profile. Other examples of cultivars either obtained in the UK and grown there as well as in Poland or obtained elsewhere but common in both countries are the pears: Conference and *Doyenne du Comice.*[56] Their Polish names are *konferencja* and *komisówka.* Despite their resemblance to English names (the second being a borrowing from French), translations such as *gruszki* Conference or *gruszki Doyenne du Comice* would not profile the right thing for the Polish mind. Fortunately, Poland and Britain have climates which most cultivars originating in one country fail to adapt to after a transfer to the other. This concerns fruits but, above all, potatoes, which, susceptible to rotting, in the UK have been geared to high precipitation rates. In the two compared languages, cultivars and breeds almost always differ at the level of conventionalisation. As to perspective, varieties tend to remain confined to specialist vocabulary but some older, culturally entrenched ones become household names not only among country people. Most Polish people know the taste or at least the lexeme of *kosztela* (an apple) or *węgierka* (a plum). Here, the comment made at the beginning of this chapter that 'specialist' does not necessarily equal 'scientific' appears particularly apt. Scientists may work towards obtaining new cultivars but farmers grow and sell them. To achieve good results, they must first see to the needs of each individually and then advertise their names to the buyers, for whom the name becomes a gate to the concept not yet stored in their minds. Here are

---

[56] See the *Book of the British countryside* (1981: 338).

some examples of cultivars in translation extracted from John Fowles' novels and bolstered by practical commentary:

1. The following refers to the cultivars of plums: 'Beauty of Bath, crisp and amber- fleshed, with their little edge of piquant acid. Still Primavera's, thinks the boy; and much better poems than bruised and wooly Pelham Widow.' Fowles (1978: 4) – 'Wspaniałe Bath, twardawe, o bursztynowym miąższu, odrobinę kwaskowate. A jest jeszcze Primavera, myśli chłopak; o ileż więcej w nich poezji niż w pomarszczonych i mechatych Pelham Widow.' Fowles (2000: 11).

The decision to preserve the English names is fully justified here. Such local varieties cannot have Polish equivalents. Alternatively, those who opt for domestication could substitute those names with the cultivars that are well-known in Poland – *węgierka* or *ulena.* That, however, diverges from the scientific truth and cannot be recommended here.

2. 'His prides there were the fruit-trees, the apples and pears some previous incumbent had gone in for and my father added to; ... I suppose people still grow them, Jargonelle and Glou Morceau, Musk Bergamot and Good Christian; the russets and pippins and wardens and codlings and nameless ones – Aunt Millie's Tree, the Yellow Devil (because it used to rot in the apple-loft), the Green Spice.' Fowles (1978: 83) – '... Jargonelle i Glou Morceau, Musk Bergamot i Good Christian, renety szare i złote, a także o imionach znanych tylko nam: Drzewko Cioci Millie, Żółte Diablę (gdyż zwykły gnić na stryszku), Zielony Cukierek.' Fowles (2000: 103).

As can be seen, some cultivar names are preserved in the original, and rightly so, since there are no Polish equivalents (except the umbrella term *Bergamoty*), some others are replaced with familiar Polish names or left out. Names such as russets, pippins, codlings and wardens cover a number of varieties, often unknown in Poland. They are not proper names, so cannot be easily retained in translation. By shape and colour some of them resemble our *malinówki* and *titówki*, pippins have an equivalent – *pepiny*, the fourth cultivar can be selected from a list of old varieties (*różanki, papierówki, kosztele, żeleźniaki*, etc. (see Dziubiak 2004). The names used only by a handful of people can be safely translated literally if they are meaningful (as above), if not, one can quote them in the original (for example, very localised Polish varieties such as *maryjki, pierzgnięta, wierzbówki, brzęczki* or *moręgi*) (see http://www.szkolkarstwo.pl/article.php?id=492-16k).

3. [Americans] declared the Cox and the Blenheim lost apples' Fowles (1978: 449) – 'Uznali, że koksy i blenheimy to jabłka do niczego.' Fowles (2000: 534). While *koksa* is officially the correct equivalent, the second variety is *Reneta Blenheimska* and though one can imagine people casually referring to the latter apples as *blenheimy*, the officially acknowledged name takes priority.
4. 'I ciągle dojrzewały nowe rodzaje [gruszek]: "cukrówki", „poczciwe", „lipówki", „sapieżanki", „bergamoty" …' (Miłosz 1989: 58) – 'And every year there was a crop of new ones, each bearing a different name: sugar pears, butter pears, honey pears, bergamots ...' (Miłosz 1981: 62).

Sugar pears are most probably identical to Polish *cukrówki*, bergamots are *bergamoty*, but the others do not seem to have any directly corresponding varieties. The names given are the names of dishes rather than cultivars. In this situation it would be advisable to preserve the Polish names.

5. The remarks above refer also to names of varieties of garden flowers:
   'Each autumn we planted her two favourite primroses round the border: Quaker's Delight and the wild oxlip, the true plant from East Anglia, not the primrose-cowslip cross.' Fowles (1978: 86) – 'Każdej jesieni obsadzaliśmy go dwoma ulubionymi gatunkami prymuli: *Quaker's Delight* i dzikiej, tej prawdziwej, ze wschodniej Anglii, a nie krzyżówki.' Fowles (2000: 107).

A word of comment, however, is due. The Polish version, quite unnecessarily, refers to Quaker's Delight and the wild oxlip as 'species', which the English text elegantly avoids. Quaker's Delight is only a variety of *primula vulgaris* (*pierwiosnek bezłodygowy*), while the wild oxlip is the name of the full species (*pierwiosnek wyniosły*).

6. Also, the translation of cultivars of potatoes follows the same advice.
   'Ben grew his own King Edwards and the rest for size and sanity, and Dan was allowed a row or two of his Catriona and Fir-

apples.' Fowles (1978: 448-449) – 'Ben sadził swe King Edwards i resztę dla wielkości i zdrowia, Dan mógł dostać rządek lub dwa dla swych Catriona lub Fir-apples'. Fowles (2000: 534).

Dormice

Dormice are the family of small tree-inhabiting rodents called *popielicowate/pilchowate* in Polish. Britain has only two species: fat and forest dormice. In Poland there are four of the thirty-four species existing in Europe, Africa and Asia: the more common, though not numerous at all, *orzesznica* and *popielica* and rare *koszatka* and *żołędnica*. In English, they are all dormice: common, fat (edible), forest, garden, respectively. They differ in colour and size. For example, the common dormouse is ginger, as is the garden dormouse except that the latter has black patches around the eyes, whereas the fat and forest dormice are grey with a whitish belly, but the latter has black-framed eyes and is smaller. The correct translation of 'dormouse' then is a) *popielicowaty*, b) *pilchowaty* or less scientifically, c) *jakaś popielica*, d) *jakiś pilch*. As we can see the level of perspective matters. And so does the way the degree of specification is manipulated with: in English it is at its lowest when the plural of the second element of each species' name is used on its own; in Polish the topmost schematicity is realised through morphological changes of the name of just one species of the entire family. It must be remembered that the ranges of distribution and population numbers of these mammals vary significantly and choosing a specific representative to stand for the whole family may be quite misleading, both in descriptions of particular regions and generalisations such as the one quoted below, where dormice receive the name of *koszatki* in the Polish version (Durrell 2002: 96) of the following text:

> Agoutis are, of course, rodents, members of that enormous and interesting family that includes creatures like the harvest mouse, which would hardly fill the bowl of a teaspoon, to capybaras that are the size of a large dog and in between these two extremes a great variety of squirrels, dormice, rats, porcupines and other unlikely beasts (Durrell 1974: 144).

Pars pro toto as well as *totum pro parte* substitution often does not work with terms, especially across language boundaries.

Dry-ground forests

In English, the meaning of the concept is expressed in the verbal form in a straightforward way: they are forests which occur on dry ground as opposed to those which grow on peat or mineral soils prone to recurring floodings. They range from conifer to broadleaved stands. The name's level of schematicity is then very high and instantiations are numerous and varied. Sometimes used in specialist papers, it sounds understandable to all native speakers. In Polish no such concept has been lexicalised. Descriptive labels such as 'lasy na suchym podłożu', 'lasy niepodmokłe', 'bory i lasy występujące na suchszych miejscach/glebach' circulate instead. The expression 'suche lasy', literally 'dry forests' ought to be approached with caution because it may signify 'forests on sandy plains and dunes', typically Cladonia pine forests, devoid of the shrub layer. The term deserves attention for one reason – translators from Polish into English tend to take it to mean *grądy* (pl.of *grąd*), the category discussed above. They reduce its profile to one type of broadleaved woodland. Moreover, they impoverish the term *grąd*, subtracting from its scope those category members which grow on damp substratum – *grądy niskie*, calling them wet-ground forest, as if denying the existence of alder swamp forests or riverside carrs, both wet indeed.

Finches

Finches are the family of *Fringillidae* (in Polish *łuszczaki*). Some of its members have 'finch' in their names, for instance, chaffinch (*zięba*), greenfinch (*dzwoniec*), goldfinch (*szczygieł*), some others have not: siskin (*czyż*), common redpoll (*czeczotka*), common crossbill (*krzyżodziób świerkowy*), etc. To make matters more complicated, a considerable number of birds within the family of *Estrildidae* and *Emberizidae* have 'finch' in the name. Their Polish names: *astryldy, papuziki, amadynki, amarantki*, etc. do not suggest their affiliation with any systematic group. Neither do any of the names of the actual family of finches. Good for transparency within one language, it turns out problematic in translation. Let us consider the following casual unprofessional observation, assuming it refers to the British reality: 'some finches flitted from tree to tree nearby'. One might rush to render it, true to facts, as: 'jakieś łuszczaki przelatywały w pobliżu z drzewa na drzewo'. But *łuszczaki* in a non-specialist context sounds odd. Wrong is the perspective and the degree of salience attached to the choice of word unlikely to be uttered in the circumstances. To return to the issue of translation – what then flitted from tree to tree? One can resort to underspecification by writing 'jakieś nieduże ptaszki …' (some quite small birds …) or overspecification: 'jakieś zięby czy coś …' (chaffinch-like birds…), where the most commonly met and recognised species represents the whole family. When we move the scene to Africa and regard the sentence included in a nature narrative, the perspective changes and *łuszczaki* no longer strikes us as incongruous: 'flocks of minute finches burst from the undergrowth and drifted across in front of us, their crimson feathering making them look like showers of sparks …' (Durrell 1975b: 68).

Grouse/game birds

In Kossak (2001: 477) the translator corrects the author rendering 'Głuszec należy do rodziny kuraków' as 'The capercaillie belongs to the order of *Galliformes*'. In the past, *kuraki* was the name of the order (*rząd*) of *Galliformes*, not the family (*rodzina*) the capercaillie belongs to. Today the Polish name of the order is *grzebiące*, the family – *głuszcowate*. In folk classification *kuraki* embraces all the birds that look and behave like domestic chickens from the capercaillie to the quail. In English they are called gallinacaeous birds or galliforms, or most colloquially, chickenlike birds. The capercaillie, black grouse, red grouse, hazel grouse, willow grouse and ptarmigan are collectively called grouse (*głuszcowate*), while those of the *Phasianidae* family (partridges, pheasants and quails) do not have a particular common name (although the *Book of the British countryside* (1981) gives them the collective name of pheasants). In Polish, *Phasianidae* is *kurowate.* While the maximal scope of the category 'grouse' numbers six members, in Britain it is prototypically associated with the red grouse, the bird for which expanses of heath are carefully maintained. On the less optimistic side, feathery predators have been decimated to boost the population size of this hunter's dream. Taking this into account, the translator from Polish must beware of using 'grouse' to refer to the black grouse (*cietrzew*) or hazel grouse (*jarząbek*), in order not to impose wrong images on the reader's conceptual base. An often encountered term *kuraki leśne* is equivalent to woodland grouse. In Great Britain, the term game birds and galliforms overlap, but in North America game birds cover a wide range of species, comprising also ducks, geese, swans, and shorebirds such as woodcock, snipe and plover. The maximal scope of the category depends here on dialect. For the translator, the awareness of the origin of receptors and thus their perspective becomes indispensable to avoid category overstretching. By writing 'game birds' in a publication meant for Americans or Canadians a much broader category is profiled than in the one for the British market. Whenever it does not adversely affect conventions of style and register, 'galliforms' should be chosen.

Lacewings

Only a comparison of the original and the translation reveals that *złotooki* and *sieciarki* have the same designatum, which in English is invariably 'lacewing-flies' (Durrell 1975a: 37 (translated as *złotooki*) and again on page 201, where they become *sieciarki* in the Polish version). To be precise, lacewings or lacewing flies embrace two families *Chrysopidae (złotooki)* and *Hemerobiidae (życiorkowate)* of the seven within the order *Neuroptera* (*siatkoskrzydłe* = *sieciarki*). Since *Neuroptera* is sometimes replaced with 'lacewings and relatives', 'lacewing-flies' can be translated as *złotooki* or *sieciarki* unless it is specified which family is meant. Although we are talking here about a scientific perspective, categorisation is as fuzzy as the folk one. Theoretically, the choice between *złotooki* and *sieciarki* is the choice between an instantiation and a schema, whose profile extends over a much larger portion of the conceptual base, but, in practice, both can be as schematic. To an amateur naturalist, not well versed in entomology, *złotook* – a familiar and pleasant sight in human abodes in autumn, stands for a specific species, *sieciarki* being perceived as a collective name for insects with cross-veined wings.

Midges

Certain midges may at first glance be taken for gnats and the English taxonomy gives this familiar name to a large number of families, while Polish names are either non-existent or alien to most people: *Ceratopogonidae* (biting midges – *kuczmany*), *Chironomidae* (true midges – *ochotkowate*), *Chaoboridae* (phantom midges – *wodzieniowate*), *Thaumaleidae* (solitary midges), *Dixidae* (meniscus midges), *Blephariceridae* (net-winged midges – *mikozkowate*), *Cecidomyiidae* (gall midges – *pryszczarki*), *Deuterophlebiidae* (mountain midges) and *Scatopsidae* (dung midges – *maraszekowate*). In other words, what has a well-known collective name in English, in Polish is segmented into a number of unfamiliar names, easily arranged in separate lexico-semantic fields. In the circumstances, the translator might be tempted to replace the strange *pryszczarka* with the incorrect but immediately recognisable *komar* or *meszka*. Whereas for the British a prototypical small biting insect nuisance is either a mosquito or a midge, for Poles it is *komar* (mosquito) or *meszka* (black fly). When a non-specialist says 'midges', he or she might just as well mean 'black flies', which, however, have a lesser degree of conceptual and linguistic conventionalisation. This overextension is determined by the abundance of lexical categories containing the word 'midge' in English. By contrast, it is extremely unlikely for *meszki* to be referred to as, for instance, *kuczmany*, even if, in fact, one has just been bitten by the latter. Expert and folk perspectives diverge, hence the translator should properly assess the level of accuracy required. The rendition of midges as *meszki* may be a sufficient level of salience in small talk in a narrative but not in a text where nature is in focus.

Sandpipers

The *Scolopacidae* family, sandpipers in English is *bekasowate* in Polish. A number of species which belong to this family bear the name of sandpiper. Some of them are *brodźce* (e.g. *brodziec pławny* – marsh sandpiper, *brodziec piskliwy* – common sandpiper, *samotnik* – green sandpiper, *łęczak* – wood sandpiper; the latter two were formerly known as *brodziec samotny* and *brodziec leśny* respectively), some others *biegusy* in Polish. On the basis of Gooders (1998), the following division can be made: *biegusy* are *Calidris* sandpipers, while *brodźce Tringa* sandpipers. Not all the *Calidris* sandpipers have 'sandpiper' in their name (e.g. dunlin – *biegus zmienny*, knot – *biegus rdzawy*, little stint – *biegus malutki*) nor do all the Tringa sandpipers (e.g. common redshank – *krwawodziób*, greenshank – *kwokacz*). Other sources use the names Calidrids or typical waders for *Calidris* sandpipers. Americans prefer to call *Tringa* sandpipers 'Shanks'. Sometimes 'Stints' functions as an umbrella term for *Calidris* sandpipers. In the jungle of various categorisations one can easily get lost. Although the one included in Gooders (1998) is quite handy from the point of view of the translator from or into Polish, it has a drawback: it comprises Latin words, which all but more advanced birders shun. To illustrate the translator's dilemmas, the ensuing fragment from *Daniel Martin* may be analysed:

> High in the darkness over the river there was a thin trisyllabic call, piercing the stasis between them. Dan said drily, "Special effects".
> "What was it?"
> "A sandpiper. Looking for somewhere warmer to live." (Fowles 1978: 604)

In the Polish text 'a sandpiper' becomes *brodziec* (Fowles 2000: 714). Technically speaking, it should be *jakiś brodziec lub biegus*, but since Dan does not choose to be very specific, putting only *brodziec* is acceptable. Of course, *(jakiś) bekasowaty* is out of the question given the circumstances of the conversation, even in the mouth of somebody like Dan, who knows a lot about nature. The low level of conventionalisation of this expression indicates a higher degree of scientific perspective. The expression's profile stretches over a larger portion of its base than the profile of *brodziec*. Conversely, in guide books, nature trail leaflets, nature albums and other didactic publications translated from Polish the boundaries around the category of *brodźce* need to be drawn clearer. In effect, *'Tringa* sandpipers' gain relative salience.

Water birds/water fowl/wild fowl

As examples of water fowl the *Book of the British countryside* (1981: 381) gives buntings, coots, grebes and moorhens. The inclusion of buntings, which can neither swim nor wade in water but live in the reeds nearby, may suggest that it is possible to put waders, for example, in this category. But then the more widely agreed term is 'wetland' birds. Consequently, the stocking of the lexico-semantic field with an exhaustive and uncontroversial list of members verges on the unattainable. Another word often used in the context of water and birds is 'wildfowl' – game water fowl: ducks, geese and swans[57]. Water birds is a category equalling in fuzziness to water fowl, where the core is occupied by the most common ducks, perhaps mallards, and at the edges of the periphery sit reed buntings, reed warblers or bearded tits, which inhabit the outskirts of lakes, ponds or watercourses. The Polish language offers similarly vague categories as equivalents: *ptaki/ptactwo wodne*, *ptaki błotne*, *ptaki wodno-błotne*. What is profiled by 'wildfowl' needs a lengthy expression in Polish, changing together with perspective. In the context of bird watching or ornithology they are *kaczkowate*. From the vantage point of hunters, fish pond owners and legislative bodies they are *wodne ptaki łowne* (game water fowl). The average layman may refer to them as *ptaki/ptactwo wodne* or by enumeration of most prototypical members of the category: *kaczki, gęsi, łabędzie*. Remarkably, these understandings contain mutually exclusive sets: only some species of ducks and geese are game in Poland, which leaves out swans, all the other species of ducks and geese and an overwhelming majority of water birds representing completely different families. The high degree of schematicity of the expression demands specification in the contexts in which underspecification may lead to the slaughter of rare and protected species.

---

[57] See e.g. Webster's, or Carton de Wiart (1950: 138) for folk understanding and http://www.staffs-ecology.org.uk/ - 8k for fully overlapping scientific usage.

3.4. Verbalised categories in context

In the previous section contexts were occasionally quoted to endorse certain claims concerning the meaning and usage of terms. The present section treats co-texts and broader contexts as a starting point. Excerpts predominantly come from novels and other narratives by authors who know a lot about nature and suffuse their works with detailed and apt observations: Gerald Durrell, John Fowles, Simona Kossak, Czesław Miłosz, Maria Rodziewiczówna. Section 3.4.2. departs from the chapter's strictures and discusses a specialist publication but the one affected by non-specialist infiltrations. We begin by analysing fragments in which terminology is bound to be obscure for the preponderance of contemporary source language receptors. Concepts other than species come first.

3.4.1. Obscure concepts and terms

Terminological obscurity may result from terms' obsoleteness or confinement to regional, dialectal or idiosyncratic use. Let us consider the following:

a) 'Za polami zaczynały się pastwiska i **pasieki**, czarna woda w koleinach pod nawisającą trawą kryła dziury, w które zapadało się po osie.' (Miłosz 1989: 35)
b) 'Raz na kupie kamieni na **pasiece** znalazł gniazdo dudków ...' (Miłosz 1989: 36)
c) 'dużo podmokłych **pasiek**' – (Miłosz 1989: 158)

For the contemporary Pole *pasieka* means 'apiary'. The archaic meaning is, according to Brückner (1996), 'the place where the forest was cut down', or by implication – open areas overgrown with herbaceous vegetation. All the three contexts eventually guide us to this meaning, but the initial reaction is to try and fit the familiar image of apiary into the depicted scene. The high level of conventionalisation of the word *pasieka* as correlative of the concept 'the place with beehives, where bees are kept for honey, wax and other bee products' makes it take precedence over less conventionalised conceptualisations. The reader expects a particular schema, automatically and subconsciously profiles one or more instantiations of it – differently arranged beehives of varying types in diverse locations. Surprised by the contextual clues, he/she looks further afield. In a) he/she learns that the word encodes something that extends over larger areas, like pastures, and unlike apiaries. B) tells them that it is something you can be 'on' – in Polish you cannot be on *pasieka* but 'in' it, whereas c) reveals that it can have wet ground – certainly not an option for an apiary. The search takes time, during which the investigator adjusts elements of the conceptual base in order to come up with the appropriate profile. In the example studied, such an effort is shared by Polish readers and the translator, English receptors being saved it. Although the words used evoke images not always compatible with the Lithuanian-Polish reality, they do not let their readers drift with their thoughts away towards bees and beehives. Here are the translations:

a) 'Past the fields came the grasslands and **moors**. Black pools in the grass-choked ruts concealed potholes deep enough for a wagon to sink up to its axle.' (Miłosz 1981: 35)
b) 'Once among a pile of rocks in the **brush**, he discovered a nest of hoopoes...' (Miłosz 1981: 36)
c) 'a lot of swampy **clearings**' (Miłosz 1981: 143)

The translator turns the unclear Polish *pasieka* into familiar 'moors' and 'brush'. While the first is simply wrong (see the discussion on *wrzosowisko*), the second is a possible but not very likely interpretation of what the logged area might look like – rough open land covered with small bushes and trees. C) offers the best translation – preserving the same level of schematicity of the modern Polish substitute for *pasieka* – *poręba* and profiling the same property – being cleared of trees.

Another example highlights problems with species identity. Although Miłosz provides us with a few, I choose *Lato leśnych ludzi* by Rodziewiczówna, which has not been translated, but abounds in mysterious, non-standard names[58]. They are presumably either coined, perhaps altered, by the people of the Polesie region in which the book is set, or misremembered by the author, who in several cases explicitly attributes a bird's name to a particular early Polish naturalist but it is not confirmed in other more reliable sources (Strutyński 1972, Urbańczyk (ed.) 1953)[59]. The reality presented by the author is alien to Polish people

[58] For example, *haty, wyraj, menteszka, kierz, lipówka, krynica, siermięga, pohybelnik, proście (prostki), płoskunie, czeczota*; and as regards names of plants and animals: *czarnygłów* (it figures neither in Urbańczyk (1953) nor in Kielak (2004)), *hryce, krzywodziób, poświerka.*

[59] The most prominent case is that of *ledwuchna* (now *lerka*), which she attributes to Jan Krzysztof Kluk (1739-1796), whereas Strutyński traces it back to Rzączyński (1664-1737), claiming that Kluk registered the bird as *dzierlatka, skowronek drzewiec, drzewczyk* (1972: 24).

nowadays, despite the fact that it continues little changed from the times the book describes – over a hundred years ago. Polesie partially belonged to Poland before the Second World War. For about 45 years after the end of the war Poles were denied entry there by the USSR authorities, and currently the Belorussian government does not welcome visitors from Poland, either. Deciphering what conceptual referent hides behind a name requires then not only the knowledge of the dialect but also familiarity with Rodziewiczówna's idiolect, which we have no key to today. The translator's progress would be undoubtedly delayed in the following fragment:

> A jakich gniazd wuj dotąd nie znalazł?
> Podróżniczka, remiza, strzyżyka i **trzciniaka świerszcza**. No – i czyża, który wedle legendy kładzie doń jakiś kamyk, co je czyni niewidzialnym (Rodziewiczówna 1998:75).

> Uncle Rosomak, whose nests haven't you found yet?
> The bluethroat's, penduline tit's, wren's and grasshopper warbler's. And the siskin's, which according to legend puts a little pebble in it, which makes it invisible (translation mine).

Trzciniak świerszcz may be what today is called ***świerszczak*** (grasshopper warbler). Although no other name seems close enough, there is a shadow of doubt, especially as Strutyński (1972) does not mention such a name at all. On the other hand, we must remember that even a very comprehensive study, such as Strutyński's, hardly ever exhausts the topic. For example, my first-hand knowledge is that the **lapwing**, officially, ***czajka***, can be locally called ***kulun*** (Warszawice, Mazovian Province), the roller, officially ***kraska***, can be ***wronka*** (Borki, Podkarpackie Province) or ***krasocha*** (Rogacze, Podlasie Province). None of these appears in the large collection of dialectal, local and old scientific names offered by Strutyński. What remains in such a situation is one's own 'informed guesswork', especially when the context does not offer any guidance. The translator should not concentrate on finding an equivalently obscure name in English, assuming that this simple book's goal is not to mystify the reader but to invite them to the world of nature and adventure. As a result, the target language reader would receive a clearly profiled category, at the high level of conventionalisation, valid from the scientific as well as folk perspective.

3.4.2. Shifts in register and imagery

In Krogulec (ed.) ducks are divided into **surface-feeding ducks** (*kaczki pływające*) and **bay ducks** (*grążyce*) (1998: 349). Bay ducks alongside **'sea ducks'** is a popular name for **diving ducks** (*kaczki nurkujące*) of the tribe *Aithyini*, family *Anatidae*. Surface-feeding ducks also have popular alternatives of their name: **dabbling ducks** or **puddle ducks** (*pławice*). Strangely enough, the authors decided to juxtapose names belonging to different registers. Condemned by language teachers, the practice is quite perplexing for foreigners, non-native speakers of English, too. Surface-feeding ducks and diving ducks, put together, form a clear contrastive pair, with the images springing to mind at once. We do not see specific species but a gestalt of a duck which paddles through water, occasionally putting its head under the surface, and a duck which from time to time disappears to pop up again. Imagery is highly schematic, verbalisations not well conventionalised. Bay ducks and sea ducks are apparently folk terms of a very high degree of schematicity – they profile uncontrastive features – surface-feeding ducks can be met at sea, especially in bays. The juxtaposition of a clearly drawn (in the contrasted aspect) specialist category and a vague non-expert one makes the reader wonder what he/she is supposed to discern as most salient. The problem is not for those who know how the boundaries of all alternative categories cut each other but for the general reader, who instinctively urged to look for sense in everything, is forced to make unnecessary effort. The juxtaposition of sea ducks and puddle ducks is potentially puzzling, too, creating a false impression that some ducks occur in puddles and others at sea. However, both categories are obscure enough for the reader to instigate an immediate search for a concept and its verbalisation with a higher degree of specificity and a more vividly marked contrast. The same applies to the *grążyce – pławice* pairing.

3.4.3. Onomatopoeia

Nature's sounds are perceived differently not only across languages but by individual listeners. Therefore verbalised categories encoding them and available for the users of the same language have extremely fuzzy boundaries. Nevertheless, conventionalised ways of entrapping non-human voices in words exist. In addition, everyone produces their own imitations, authors excelling in creative onomatopoeic transcriptions. In principle, most such endeavours should be judged successful, but even the staunchest relativist would notice that sometimes their outcome does not resemble the prototype in the slightest. The translation of 'buczący samiec [bąka]' as a humming male (Sawicki 2003: 38-39) or 'odzywające się samce [derkacza]' as humming males (ibid.: 42-43) is ludicrous. The bittern (*bąk*) booms in English, the corncrake (*derkacz*) utters harsh, grating *crek-crek-crek* (see Gooders 1998: 411). Both sounds are loud, alarming and regularly broken, unlike the gentle continuous humming of bees. On the other hand, one can never expect a perfect correspondence of notation, appreciating the approximation of auditory imagery, such as in the text extracts below:

> the **harsh rattling cries** of the guans, the **incredibly loud trumpeting song** of the seriemas, the **chittering** of the coatimundis, and an occasional **dull rumble, as of distant thunder**, from the puma … there was a **steady scrunching noise** that came from the agouti cage, for it was always engaged in trying to do alterations to its living quarters with its chisel-like teeth. (Durrell 1974: 157)

> **chrapliwe nawoływania** penelop, **niesłychanie donośne trąbienie**, jakie wydawaly kariamy, **szczebiot** koati oraz **przypominające odległe grzmoty warczenie** pumy … rozlegał się **monotonny chrobot** – to samica aguti usiłowała, jak zwykle, przemeblować mieszkanie za pomocą pilnikowatych zębów. (Durrell 2002: 105)

The display song of the capercailie (*głuszec*) has always excited naturalists, hunters and inspired writers not because of its melody but for its extraordinariness and mysteriousness, epitomising calls of a primeval forest. No wonder both languages have established verbalisations of the song: 'the lekking song [of the capercaillie] consists of four phases: tapping, drum-rolling, cork popping and gurgling-wheezing' (Zawadzcy 2003: 142). This closely corresponds to the description in Gibbons (1991: 145). The only, but significant, difference is that the first stage is referred to as 'clicking'. In Polish tradition, the song has either three stages: *kłapanie*, *korkowanie* and *szlifowanie*, or four – with one more distinguished after *kłapanie* and called *trelowanie,* which is basically an accelerated version of *kłapanie*[60]. Less conventional transcriptions, less schematic, profiling slightly different aural impressions can be found, too:

> The song, praised so much by poetically-minded hunters, is a series of short, 7 to nine-second **clatters, rattles, puffs, rustles, rasps and pops**

[60] See Skoczylas (1990), Sokołowski (1992), Matuszewski and Morow (1994) or Jonsson (1998).

> like those of a cork coming out of a bottle. Unfortunately, while rasping, the normally alert bird stops hearing for a moment … (Kossak 2001: 478).

In Polish, the sounds are rendered as ***klekoty, terkoty, pyknięcia, szelesty, zgrzyty*** and ***puknięcia***, respectively. 'While **rasping**' is 'podczas **zgrzytania** (tzw. **szlifowania**)'.

And one more:

> *Tek-ap*…. Faster and faster, until the **hooting** was joined by another sound, rather like **knife-grinding**; that's when Romuald took his first hop, followed by a second, turning stock-still on the third. Thomas was afraid to budge before he was sure of his timing. But the instant the grouse began a new series, he was ready; taking his cue from the **whetting**, he jumped together with Romuald. One, two, three. He knew why the interval: the bird was momentarily deaf. They could risk a clamor now, so long as the body was prepared to become a statue on a moment's notice. (approaching the capercaillie on a lek) (Miłosz 1981: 174-175).

The original (Miłosz 1989: 158-159) contains the same 'tek-ap' notation of praiseworthy accuracy. The word 'hooting' penetrates into the description on the translator's own initiative and in fact profiles the prototypical call of owls. A rapid series of clicks/taps can scarcely be called 'hooting'. The translator's manipulation with the profile fails miserably. By contrast, 'whetting' (in Polish *szlifowanie/ostrzenie*) seems a well-adjusted addition to the scope of the category of capecaillie's vocal repertoire in English.

3.4.4. Puns and other manipulations

It happens, especially in Fowles' novels, that nature does not just reside in descriptions, constituting a backround but becomes a protagonist or the fathomless inventory of associations. One of the most refined forms in which it enters the plot is through puns. While the translator's task amounts then to finding associative correspondences most similar to those in the original, the preservation of referential verisimilitude of chosen concepts is paramount. A closer look at an excerpt from *Daniel Martin* should clarify our point. Daniel is a boy in England in wartime when he describes the following:

> Now a plane drones slowly over, …. A **Tiger Moth**. Another name. I also know the real (though do not know that in the unconscious "real" my redeemer cometh) tiger moth: the fluttery, zigzag-striped, chocolate-and-

> cream, black and red-orange **Jersey Tiger**. We catch some every year in the garden (Fowles 1978: 91).

> Nad głową buczy, lecąc powoli, samolot … . Błyska myśl: "**Tygrysica**". Każdego roku łapiemy kilka w ogrodzie (Fowles 2000: 113).

The name of the plane is a culture-specific issue. The associations the boy has, would not be evoked in the Polish mind for many reasons. Firstly, names of planes are usually left in the original, sometimes slightly adapted, secondly, the name of the moths is completely different. Tiger moths are **niedźwiedziówki** (which derives from 'bear' not 'tiger'), and Jersey Tiger is **krasopani hera** (no reference to any dangerous animal). Less baffling for seekers of comparisons between civilisational achievements and nature would probably be playing with the word *niedźwiedziówka* than *tygrysica*. Alternatively, after a straightforward operation, the text could look as follows:

> … Tiger Moth, czyli *niedźwiedziówka*. Znam także prawdziwą (chociaż nie wiem, że w owej nieświadomej „prawdziwości" jest moje zbawienie) niedźwiedziówkę: trzepocącą, z zygzakami na skrzydłach, kremowoczekoladową, czarną i czerwonopomarańczową, zwaną krasopani hera. Każdego roku łapiemy kilka w ogrodzie.

The pun is gone but at least the reader does not imagine the unlikely figure of a female tiger (*tygrysica*) caught every year in Daniel's garden. In English, two facets gain salience simultaneously: a clever wordplay inducing aesthetic pleasure and Daniel's ability to see correlations between man-made and natural objects and phenomena. In Polish, only the second one is instantiated and, actually, boosted, because in the Polish version Dan makes a strong impression as highly expert at entomology. *Krasopani hera* sounds alien to almost every Pole and does not evoke associations with anything familiar in nature. Although in both languages the degree of specialist perspective is high, in Polish it further increases. Overspecification of one aspect may be regarded as a loss but relative to the existing official translation, it seems a step ahead.

A similar loss seems inevitable in the case of the following pun:

> [The hoopoe utters] głuchy, dobiegający jakby z głębokiej studni głos, namawiający nas, żebyśmy kogoś upupili: „upuup ... upupup"

> voices as if from a deep well: 'oopoop ... oopoopoop ...'. In the Polish language they are somewhat akin to a phrase meaning 'make a fool of …' (Chylarecki and Sawicki 2003: 90-91).

We are offered an explanation, and in the context of a publication in which facts play a far more important role than toying with language it ought to be standard procedure. Admittingly, the enhancement of specialist knowledge and failing to give salience to the message that nature may inspire cultural sensitivity causes the passage to become less multi-faceted and, to be quite plain, less light-hearted.

The last example provokes a reflection on creating imagery detached from conventionalised verbal categorisations in the target language, which criss-cross frontiers between the specialist and folk usage. The overall effect of this kind of translation does not strike us as clumsy or projecting the impossible but components of the image differ in two languages both as far as their quantity and quality are concerned. Let us compare the original text excerpted from *Daniel Martin* again with its translation.

> She [Nancy] began picking centaury and eyebright there, then knelt by a clump of the pink-headed starry flowers.' Then Daniel said to Nancy "They are not much good, actually. They won't open indoors." Nancy: "My dad likes'em" . D: "They are called centaury". N: "Earthgirls" … That's what we say".
> …He didn't find that quaint (**or inexact, she meant earthgalls**); but embarrassing. His intellectual superiority, he was so anxious not to seem stuck up, he shouldn't have showed off about the real name … (1978: 379-380).

> Zaczęła zbierać centurie i świetliki, potem przykucnęła przy kępce różowogłowych gwiaździstych kwiatków.' ... D: Mało z nich pożytku. W domu się nie rozwiną. N: Mój tata je lubi.' D: Nazywają się **centurie**. N: **Panienki**. ... Tak mówią u nas.
> Nie poczuł dumy, a bardziej zakłopotanie. Za wszelką cenę starał się nie akcentować swojej intelektualnej wyższości, nie może jej poprawiać ... bał się, że ją urazi.' (2000: 455)

The problem is how to render the folk name of the centaury, whose scientific name in Polish is *centuria.* The most common folk name is *tysiącznik* (derived from *tysiąc* = a thousand). It comes from a mistranslation of the Latin *centum*, which is *sto* (a hundred). It seems possible to construct a dialogue on the basis of this mismatch between the folk name and the meaning of the scientific name. Nevertheless, the translator has chosen a literal translation, creating reality (calling *tysiączniki – panienki*) rather than referring to the existing one. He has also omitted the passage in which Daniel realises that the folk name used by Nancy is mispronounced. One figure is removed from the image, so the figure/ground alignment loses the sharpness of contrast. In the source text

Daniel not only knows the scientific name but also the folk one, which Nancy must have once misheard and since then used incorrectly. The greater is his embarrassment and the reader's awareness of it. If *tysiączniki* took the place of *panienki,* the mispronounced version could be *tysiączki* – quite a credible corruption in the mouth of naïve categorisers. The missing bracketed comment '(or inexact, she meant earthgalls)' could become '(albo nieprecyzyjne, powinna powiedzieć tysiączniki)'. To put it briefly, the difference between the source and target imagery is that of tinge rather than radically contrastive hues.

3.5. Recapitulation

This chapter has strived to show links of conceptual and linguistic categorisation with imagery in its sensory, visual and auditory aspects. It has combined simplified cognitive theoretical analysis with practical advice for translators, who encounter nature terminology in texts popularising nature as well as in literary works. The elucidation of the intricacies of categorial demarcations, shifts and membership has involved a broad spectrum of data – linguistic, cultural, environmental, synchronic and diachronic alike. Lexical items examined have been treated as complex categories in due regard to all conceivable (to the author) conventionally sanctioned usages with the stipulation that:

> Even when all its [= complex category's] attested values are plausibly analyzed as instantiations of a single abstract schema, or as extensions from a single prototype, there is no way to predict from the schema or prototype alone precisely which array of instantiations or extensions – out of all the conceivable ones – happen to be conventionally exploited within the speech community. (Langacker 1987: 370)

By and large, the study of errors has been eschewed, though faulty translations have sparingly been cited to enhance correct usages. In the next chapter, conversely, mistranslation comes under the spotlight.

CHAPTER FOUR

A DISTORTED VIEW OF THE WORLD: A MISTRANSLATED CATEGORISATION OF NATURE

4.1. Categories of translation errors – an introduction

In this chapter I would first like to see in what measure the translation errors relating to the functioning of nature terminology in texts can be classified with the help of a multi-faceted classification arrived at in Hejwowski (2004a: 198-234). The author's proviso that the categories he has distinguished may overlap and 'it may be difficult to decide how to classify a given example' (2004a: 234) acts as a stimulus for me to conduct a more reductive, though still sensitive analysis. I thus undertake an attempt to present errors in terms of undesirable changes in imagery (in all or some of its different dimensions) evoked by the wrong identification of a category prototype or a serious departure from the one statistically[61] best recognised. Alternatively, we can talk about an abuse of such features of categories as stretchability or fuzziness with regard to a particular background framing. Overstretched and overfuzzy categories lead to the visualisation of a wrong image, the whole series of which contributes to the distorted view of the world. Responsibility for the distortion often rests with translators, who try to reconstruct in the target language a linguistic view/picture of the world presented in the source language. Prior to finding the right words and grammatical structures they must envisage corresponding concepts – conceptual categories, establishing their prototypes and adjusting their boundaries. Such activities consist in taking countless, risky decisions, requiring versatile knowledge and, not

[61] In the sense Langacker uses the word, writing that the prototypical model 'is founded on the prediction that entities are more readily accepted as members of a category the more closely they approximate the prototype – but its predictions are statistical rather than absolute' (Langacker 1987: 49).

infrequently, betting on the best of too many good solutions or on the most palatable of a number of scarcely satisfactory ones.

Translators, rarely specialists in natural sciences, may be perplexed by the overabundance of names and peculiarities connected with their use. They often are, despite a generally valid idea that it is easier for the translator to master intricacies of specialist knowledge than for the specialist to acquire top-quality translation skills (advocated by Hejwowski 2004b: 157). Faced with some data presented in the previous chapter, translators may draw wrong conclusions and make inadequate choices. For instance, it is not easy to internalise that in some circumstances *jaworzyny* (sycamore forests) should be rendered as beech forests. The discovery that sometimes fens, essentially distinct from bogs, can be classified as quaking bogs, is likely to lead to an appalling assumption that whether you translate *bagno/torfowisko* as 'fen' or 'bog' does not matter so much. A similar effect can be brought about by the Natura 2000 idea that 'bog woodland' accommodates birch-pine **fen** carrs. An unwelcome overextension on other contexts is almost bound to happen without due diligence. Having learnt that riverine/riparian/riverside forest is *łęg* in Polish, we may be tempted to automatically translate it as one of these, overlooking the fact that it can also grow in places fed by flowing groundwater rather than surface water of rivers and streams. Tired of guessing when and where *grąd* is 'oak-hornbeam-linden forest', 'oak-hornbeam-linden-ash forest', 'oak-hornbeam-spruce forest', 'oak-hornbeam-pine forest', 'oak-hornbeam forest', 'hornbeam forest' or 'beech forest', the translator may resort to the "blanket" 'oak-hornbeam forest', which in a particular case may diverge from the truth. Equally undesirable is an uncritical approach to the taxonomical nomenclature of plants and animals, manifested in using obsolete or local names, or the least frequent of officially approved alternatives, which authors sometimes indulge in.

Beside the impact of the unstable terminological scene the translator's decisions are also motivated by the extent to which s/he has absorbed the cultural background of the source and target languages. Low levels of absorption give rise to translation errors committed both through conscious manipulations and hasty subconscious decisions. As it emerges from the corpus of data compiled for this book, in relation to texts in which nature terminology occupies a prominent position such errors typically involve:

- recognition of meaning (polysemy, synonymy, inference from context);
- narrowing meaning;

- overextension of meaning;
- misnaming or unnecessary renaming of culture-specific organisations, institutions, forms of nature conservation, habitats, landscape features, etc.;
- names of species;
- names of higher taxa (groups of species);
- confusion of American English and British English nomenclature;
- omitting fragments of text or single words;
- adding fragments of text or single words;
- changing onomatopoeia and descriptions of sounds;
- failure to correct mistakes made by the author of a non-literary text;
- adjusting the text to presumed expectations of projected readers (style, domestication).

In most cases, the criterion for considering something an error is a drastic departure from the most up-to-date scientifically agreed prototypes, engendering the incompatibility of imagery. With regard to the omission and addition of fragments of text or single words as well as the changing of onomatopoeia and sound descriptions the criterial rigidity is barely possible. The translation critic ought to allow for counter-arguments validating solutions s/he rejects as infelicitous, which does not deprive him/her of the right to judge a particular omission/addition/change as erroneous.

Since this study, *inter alia*, aims to offer practical guidelines, nearly the entire database of translation errors that has been compiled is arranged under appropriate categories of Hejwowski's taxonomy. However, some of the wrong species names have been omitted as they fall into the class already richest in membership. These are usually the misnomers which can hardly be attributed to causes other than the translator's sloppiness: porpoise instead of tortoise, *czapla* (heron/egret) instead of *czajka* (lapwing), or turning the original's chiffchaffs and willow warblers into *pliszki* (wagtails) and *strzyżyki* (wrens). In the later imagery-based analysis, I only reach for selected examples to demonstrate the method's comprehensiveness. While I quote original texts with all the errors that they contain, corrections are suggested only for those mistakes that pertain to the distorted view of the natural world. The remaining ones, however glaring they may be, are not pointed out in order to keep the argumentation focused. For the sake of easy reference throughout the chapter, Hejwowski's classification is recalled below in full:

I. errors of syntagmatic surface translation
1. dictionary equivalents
2. false friends
3. commonly accepted equivalents
4. calques
5. unjustified borrowings

II. errors of mistaken interpretation
1. misreading syntagms and wrong interpretation of verb frames
2. misinterpreting scenes and scripts
3. misreading the text's modality

III. realisation errors
1. target language errors
2. wrong evaluation of the recipients knowledge
3. insufficient knowledge of the subject-matter

IV. meta-translation errors
1. choice of translation techniques
2. additions
3. omissions
4. two versions
5. too many or too few footnotes
6. adopting a wrong translation strategy
7. corrections
8. changing the text's intertextuality

Tentatively, I assume that some of the categories will apply to my examination to a lesser extent, as it focuses on nature terminology in context rather than narrative strategies, literary allusions or idiosyncracies of style. Consequently, they may not be exemplified as abundantly as others, which, considering the relative richness of the database, hopefully, indicates problem areas to be tackled most conscientiously.

While assigning categories of Hejwowski's classification to examples, I do not use Langacker's terminology, concentrating on practical advice, formulated in the language which is not theory-laden. It is only in 4.3. that I embark on employing the cognitivist (*inter alia* Langackerian) terms of which I have taken advantage in the preceding chapters.

4.2. The distorted view of the world in Hejwowski's classification of translation errors

4.2.1. Errors of syntagmatic surface translation

Syntagmatic translation occurs at the level of words, phrases and sentence structures (called syntagms by Hejwowski), correspondences between which may deceptively seem to be readily established in the source and target languages, without reading deeply into the text, but on the basis of relevant dictionaries.

4.2.1.1. Dictionary equivalents

It happens that translators pick the most common dictionary equivalent without looking up the actual one imposed by a given context, for example:

Wrzosowisko

In Kłosowscy 'rezerwat Bielawskie Błota. Wrzosowisko' becomes 'Bielawskie Błota Reservation. Moorland' (1999: 143). Dictionaries usually tell us that *wrzosowisko* is 'heath, moor, moorland'. They fail to note that moorland, as discussed here earlier, is upland overgrown with heather, gorse, grasses and lichens. The Bielawskie Błota Nature Reserve (located north-west of Gdańsk) protects lowland peatland with heather. In order to preserve the brevity of the caption, 'moorland' should be replaced with 'heath'. The same remark concerns the following quote, 'Zachowanie ukrytych tam polanek, wrzosowisk i mokradeł …' (ibid.: 98) – The sentence postulates the preservation of glades, heaths and wetlands in Puszcza Biała and Puszcza Kurpiowska, both forests growing on the flat plain of Mazovia. 'Moorland' instead of 'heath' in the English version is again out of place. Kłosowscy (2005) includes similar mistranslations: '[the nightjar's] environment are dry pine forests intersected by dunes, logged clearings and moorlands' – again, this is to stand for *wrzosowiska* (204) and 'One of the prettiest moorlands in Poland covers the open dunes and burned out clearings of the Kampinoski National Park' (ibid.: 233). To make it clear, the park lies in a vast lowland and surely does not even distantly resemble the moors of Yorkshire or the Pennines. According to the translator of Karolczuk-Kędzierska (2007: 118), *wrzosowisko* near Lake Łebsko is moorland, though for the same reasons as above it is apparently a heath. Also the English version of Miłosz (1981: 171) repaints Lithuanian *wrzosowiska* – heaths as moorland and moor.

Kępka/dolinka

In Różycki and Iwaniuk *struktura kępkowo-dolinkowa* of the alder carr is mistranslated as 'clump-valley [structure]' (2001: 4), which betrays the translator's thoughtlessness. The most common dictionary equivalents cannot be used in this specialist context. Instead 'hummock-hollow' must be employed, which dictionaries hardly ever list under *kępka* and *dolinka.* For 'hummock' they typically have 'wzgórek, pagórek, wzniesienie', for 'hollow' – 'wydrążenie, dziupla, dziura' and only further down 'wklęsłość terenu, kotlina, dolina'. A similar problem appears in Sobociński: 'W rejonach podtopionych przez wodę gruntową lub w dolinach rzecznych spotykamy olsy i łęgi. W pierwszym z tych typów lasu, olsze rosną na wysokich, często porośniętych bujnymi mchami **kępach**.' (2004: 7) – 'In the wetlands and along river valleys we shall find alder carr and riverside carr. In the first type of woods the alder grows on high **holms** that are often covered by thick moss.' (ibid.: 18). 'Holm' is chiefly used in British English and stands for 'a small island in a river or lake, or near the mainland' and 'low flat land near a river' (Webster's). In fact, in Polish the word matching the definition is **kępa**, the one also appearing in the context of the alder swamp forest. Nevertheless, the English language for the raised patches of ground in the alder swamp forest has, as already pointed out, a different word – **hummocks**. This example can perhaps be better classified as a mistaken syntagm because 'holm', unlike 'clump' is not the commonest of equivalents.

Rushes/*Szuwary*

Lewandowska and Lewandowski (eds) write '... obecność znacznych powierzchniowo torfowisk – mszarów i **szuwarów** ...' – '... the presence of large peatbogs – morasses and **sedge-rushes**' (1999: 5 and 7). The translator has avoided the mistake of translating ***szuwary*** in this context as **reeds**, probably realising that their occurrence in a peatbog is undesirable rather than being a factor worth advertising in a guidebook published by the landscape park authorities. He or she (unknown) has picked the most common dictionary equivalent. Unfortunately, the word 'rushes', used together with sedges, is inaccurate. It refers to aquatic communities of such plants as bulrush or common club-rush. Certainly, the peatlands are covered with ***sedge communities*** or are ***sedge beds***.

4.2.1.1.1. Most common though incorrect dictionary equivalents

At this point it is necessary to complete Hejwowski's type under discussion with a subcategory of most common dictionary equivalents which persist despite being obviously inadequate.

Zagajnik/gaik

Dictionaries obstinately present them as coppice and copse rather than 'young-growth stand', 'small immature wood' and other ways tailored to a particular context. The following comes from Iwaszkiewicz: 'sosnowego gaiku: tych kilku wysokich sosen, które stały za ogrodem na wzgórzu' (1987: 440) – 'the pine coppice – those few tall pines that stood on higher ground at the bottom of the garden' (2002: 63). There are two problems to be discussed. Firstly, in the Polish version, the pines grow on a hill behind the garden. It is difficult to imagine 'higher ground' at the bottom of anything. Secondly, do a few tall pines form a coppice? Despite conflicting definitions (see Chapter 3.3), a group of a few tall pine trees, by implication without a thick undergrowth, is neither a coppice nor a copse. 'Grove' would definitely adhere to the original image more faithfully.

Torfowiska

Dictionaries keep promoting 'peat bogs' as an umbrella term for all peatland/mire types. As explained in Chapter Two, this is grossly inappropriate since it signifies but one type – 'raised bogs'. Translators disseminate the error still farther. They are consistent in it in Karolczuk-Kędzierska (2007), Łapińscy (2007), Sidło, Błaszkowska and Chylarecki (eds) (2004), Kossak (2001), Kot and Dombrowski (eds) (2001), Sawicki (2003a and 2003b), Lewandowska and Lewandowski (eds) (1999), Polska. Mapa ochrony przyrody 1:750000. (2000). Occasionally one can find it in Pudlis (2003), Kłosowscy (2005), Budrewicz (2002) and a number of other publications. Sometimes, probably by extension, 'bogs' unjustifiably plays the role of the hypernym for all mire types. Let us consider a fragment from Kaźmierczakowa and Zarzycki (eds). First we read: 'the plant [fen orchid] is connected with **fens** ...' and a little below: 'The main threats to the species are drainage of **bogs**, ...' (2001: 575). In view of the first statement, the second is illogical or even absurd. Similarly, on page 544 we are informed that the musk orchid occurs in Poland in one site only, i.e. the Rospuda Valley, where it grows in an extensive **bog**. The

Rospuda valley comprises some very small patches of raised bogs. The adjective 'extensive' can only be attributed to its **fens**. However, sometimes the context fully justifies the use of 'bogs' as a generalisation, for instance, on page 498, where the following appears: 'forest bogs overgrown with old stands, largely composed of spruce'. Here the probability that the spruce stands have developed on bogs rather than fens is very high. Page 424, in turn, exemplifies the correct employment of 'fen' as a generalisation. We read there about the trampling of fens near Mount Pilsko. The Polish text uses the general term *torfowiska* at that point but earlier makes it clear that the area is indeed a fen. Using the specific instead of the general is then fully justified. But especially inexperienced translators must remember that 'mire' or 'peatland' are much safer as umbrella terms. Almost identical instances of the misapplication of 'bog/bogs' can be found in Bzura and Scherbuk (2007). Some other specific contexts are quoted under the subcategory of realisation errors called 'insufficient knowledge of the subject-matter' (c.f. page 183).

Moorhen

Let us consider three fragments of translated texts. The first comes from *Daniel Martin*, where as part of the description of wetlands near Oxford the following sentence appears: 'Somewhere further up the cut a **moorhen** croaks.' (Fowles 1978: 24) – 'Gdzieś znad kanału dobiega skrzek **pardwy**' (Fowles 2000: 34). The second is excerpted from *The zoo in my luggage – Zoo w walizce*: 'we would sit in a row above the weir, watching the **moorhens** paddling by ...' (Durrell 1975b: 179) – 'siadywaliśmy więc rządkiem nad tamą, obserwując brodzące w wodzie **pardwy** ...' (Durrell 1991: 269).

In both 'moorhen' is rendered as *pardwa. The great English-Polish dictionary* (Stanisławski 1990) and, surprisingly, *The great English-Polish dictionary* (Linde-Usiekniewicz (ed.) 2004), where a lot of mistakes found in the former were corrected, list exactly the same pair as equivalent to each other. The truth, however, is different. *Pardwa* is 'willow grouse', *pardwa szkocka* – the prototypical *pardwa* for the British is 'red grouse'. Both are land-living birds akin to domestic chickens and unable to swim. Their typical habitat are heather moorland and bogs. Feeding on bog vegetation, they avoid places of open water. The moorhen inhabits water bodies and the surrounding wetlands. It can swim, dive and walk on water vegetation – which indicates how light it is, much smaller and lighter than any grouse. The image of the moorhen 'paddling by' or croaking in the reeds is true-to-life, whereas the image of any of the grouse family doing the same belongs to the world of fantasy, and as a depiction of reality, is grotesque. To conclude, translators are better advised to rely on specialist literature, where they can learn that 'moorhen' is ***kokoszka*** (once known as *kurka wodna*), rather than general language dictionaries, because even very good ones contain errors leading to absurdities such as the one described above. It turns out that the reliance on a dictionary falsification can blot out an ostensive indication of a specimen – as a TV programme (Wildlife Aid. Animal Planet. Shown 22nd Nov 2007) illustrates. On the screen we see a moorhen trapped in a sludge receptacle. A specialist is trying to get it out with the help of a pole with a net tied to it at the end. There are at least two other people around. One of them is translated as saying that the bird is *pardwa*. Later, after the operation is finished, we can hear: 'to nie pardwa, to łyska'. Now the creature might even be taken for a coot (*łyska*) because it is all smeared in black sludge and the distinguishing features such as a white beak and forehead are not visible. But before getting dirty it was patently obvious that the species was *kokoszka* not *łyska*. *Pardwa* is out of the question. Nothing points to this species except the entries in Linde-Usiekniewicz and Stanisławski's dictionaries.

4.2.1.2. False friends

According to Hejwowski, false friends are not only 'absolute pairs' like eventual – *ewentualny*, 'but also pairs of similar[-looking] words whose meanings in the two languages in question are only partly similar' (2004a: 204). One such pair is occupation – *okupacja*. The pairs discussed below belong to the latter type. In one of their senses they overlap, but not in the one imposed by the context, which has escaped the translator's attention. They may overlap with the previous category because bilingual dictionaries do give wrong equivalents of such words.

Orchids – *Orchidee*

In *The ebony tower* we find the following fragment:

> "Oh Kate, look! There are my butterfly orchids". And Annabel leads the way up a little sunlit clearing in the trees beside the path, to where five or six slender, white columns of the delicate flower stand from the grass ... "Why are they yours?" "Because I found them last year." ... "They're wild orchids", says Annabel. "Butterfly orchids". (Fowles 1975: 242)

At that the other participants of the outing come and

> [t]hey seem disappointed, a little at a loss when they see how small and insignificant the plants are. (ibid.: 242)

In the Polish version, Annabel exclaims: 'Kate, patrz! Moje dzikie orchidee!' (Fowles 1994: 260). The translator must have been mesmerised by the word 'orchids', which looks almost like Polish *orchidee.* The foreign-sounding *orchidee*, though, stands for non-native flowers known from rainforests and as pot plants. The wild European species are called *storczyki*, the name having a familiar, 'homely' flavour.

Grape hyacinths – *hiacynty*

Grape hyacinths mentioned in Durrell (1976: 189) appear in the Polish version as ***hiacynty*** instead of ***szafirki*** (Durrell 1999: 190). Alternatively, it can be treated as a meta-translation error, in which a hypernym replaces the lower-level verbalised category.

4.2.1.3. Commonly accepted equivalents

Hejwowski exemplifies them, *inter alia*, with the word 'officials', whose commonly accepted equivalent is *urzędnicy* ('clerks'), although in most texts it should be translated as 'przedstawiciele rządu/ministerstwa'. As regards nature terminology, two words immediately spring to mind: *grąd* and *puszcza*. The first one tends to be indiscriminately rendered as oak-hornbeam forest in spite of the existence of a great variety of combinations of tree components (see Chapter Two and Three). The second automatically becomes 'primeval forest', even if the forest in question has been long managed by humans, preserving *puszcza* in its name but not in its pristine character. Not to mislead English receptors, such woodlands ought to be called 'forests', which already suggests that they may be expansive and mature but not necessarily left untouched from time immemorial. In Kłosowscy, a photograph showing an adult **pygmy owl** has the caption: 'An **owlet**' (1999: 130), suggesting it is a young owl. The translator must have been misled by the Polish ***sóweczka***, which can mean both the name of the species 'pygmy owl' and function as a diminutive. Adult pygmy owls are small and cute, so reaching unreflectively for the most commonly accepted equivalent rather than the species name proved irresistible.

4.2.1.4. Calques

In the case of nature terminology, calques abound in names of species. They are enumerated beneath under alphabetically arranged source authors.

I. Chylarecki and Sawicki (2003):

How deeply ingrained a category name can be in the mind of the translator – a native speaker of the source language – is visible in confrontation with a debatable but widely agreed name in the target language:

> 'Ze zwierząt preferujących środowisko ziemnowodne spotykamy tu jeszcze ... i karczownika ...' (40) –

> 'Of the species noticed by their preference for land-water environment we can meet also … a **water vole** (a **land-water** one of course) …' (41)

The full Polish name of the water vole is ***karczownik ziemnowodny,*** which reflects the animal's way of life because it needs both land and water to live. The English name happens to reflect only one aspect of its existence but unambiguously refers to this particular species. The translator seems to forget that the readers of the English version do not feel a kind of discomfort reading the name as he does, knowing Polish, and can even consider the bracketed remark patronising or at least fuzzy.

II. Chylarecki and Sawicki (2003):

> '**Piskliwiec** – bo tak go się skrótowo nazywa (58) –

> ' "**Squeaker**" – as they are sometimes called' (59).

'They' (why not 'it'?) refers to the common sandpiper. While it is true that Polish birders often use the name *piskliwiec,* the English have for the word 'squeaker' a lot of other applications but not as an alternative name of this bird, the more distant being, for instance, 'spy working for the police', the closer but therefore more confusing, 'young race pigeon', etc. Writing 'as it is sometimes called by Polish birders' would have at least put on the right track those foreign readers who thought they had just learned a new sense of the word 'squeaker'. By no means would it make the translation correct in its entirety, though (see 4.3.1.).

III. Fowles (1977: 221):

a) 'I thought I heard a **sub-alpine warbler'** – '... zdawało mi się, że słyszę głos **podalpejskiej gajówki**' (Fowles 1982: 264). Today the second element is properly translated as ***pokrzewka*** but in Wetmore's classification of birds (from 1960), never vehemently propagated in Poland as it stood, warblers were *gajówki*. The bird is called ***pokrzewka wąsata***. Giving a self-coined name is yet another example of underestimating the reader's eagerness to pinpoint the actual referent. The assumption is that s/he will be satisfied with a scientifically-sounding name, however impossible to trace back to the real creature.
b) The following is an example of a calque combined with mistaking two source language syntagms:

> Just as he passed us I heard high overhead one of the frequent sounds of the river, a very beautiful one, like the calling of Tutankhamen's trumpets. The flight cry of the **black-throated diver**.' (Fowles 1977: 305)

> Kiedy nas mijał, usłyszałem wysoko nad głową jeden z tych nadrzecznych ptasich krzyków, cudowny dźwięk nawoływania trąb Tutenchamona. To krzyczał w locie **nurek o czarnej piersi**. (Fowles 1982: 364)

In this fragment a bird lover is sharing the delight he derives from hearing the outstanding voice of the concrete species, the voice many birders long to hear in the wild. It is uttered by the black-throated diver – ***nur czarnoszyi***. The translator calqued 'diver' – *nurek*, which in Polish means 'somebody who dives' not *nur*. Subsequently, she mistook *szyja* for 'breast' and arrived at ***nurek o czarnej piersi*** (black-breasted diver), which sounds ridiculous and is almost an offence. A Polish naturalist, unaware of who has made the error might even think the author himself or his character are incompetent and in reality do not care much about the natural world.

IV. Fowles (1978: 351-352):

> The two ravens' calls became more frequent and I saw them attacking **a** red-tailed hawk – … Though the Americans term the bird a hawk, it is in fact a mere racial variant of the English buzzard, **indeed indistinguishable at a distance**, just as the mobbing behaviour was indistinguishable from similar scenes in any South Devon sky;

> Okrzyki kruków stały się częstsze i zobaczyłem, że atakują **jastrzębia o czerwonym ogonie** – ... Chociaż w Ameryce ptaka tego nazywało się jastrzębiem, była to w istocie odmiana angielskiego myszołowa, z daleka **równie trudna do odróżnienia**, podobnie jak walka była nie do odróżnienia od podobnych scen rozgrywających się na niebie południowego Devonu, ... (Fowles 2000: 422).

The Polish name of the bird is *myszołów rdzawosterny*. The narrator, well-acquainted with bird silhouettes and scientific names of species is unlikely to say he can see 'jastrzębia o czerwonym ogonie' (literally, 'a hawk with a red tail'), especially as he then proceeds to explain that the American name is inadequate. The English **'indeed indistinguishable at a distance'** refers to the red-tailed hawk being indistinguishable at a distance from the buzzard which occurs in Britain (the common buzzard), which is supposed to confirm their close relationship. The fragment **'równie trudna do odróżnienia' – 'equally indistinguishable'** together with the descriptive rather than scientific name of the bird in question at the beginning might suggest that the English buzzard and red-tailed hawk are equally difficult to tell apart from a hawk, which of course is not what is being claimed.

V. Fowles (1978: 542):

'**[S]pur-winged plover**' is calqued as ***ostroskrzydła siejka*** (Fowles 2000: 643), which does not exist in the Polish nomenclature. It is just a literal translation of the English equivalent, with the second element misspelled ('plover' is *siewka*). The real Polish name is ***czajka szponiasta*** (in the past – *skrzydłoszponka brunatnoszkrzydła*).

VI. Kłosowscy (1999):

a) ***dubelt*** – '**double snipe**' (50)[62] instead of 'the **great snipe**'
b) ***orlica*** – **eagle fern** (156), which is a half-calque in which 'eagleness' of the Polish name is preserved but not literally – *orlica* means 'female eagle'. The name used widely is '**bracken**' – in Blamey, Fitter, and Fitter (2003) the only one given, in Anioł-Kwiatkowska (2003), strangely, the first option is 'adderpid', the second 'western bracken'.

[62] If only one page number is given, it means that both the original and the translation appear on the same page in the same book.

c) the bird in the photograph is presented as 'A **large curlew**' (200), which suggests it is a particularly big specimen of a curlew'. The Polish ***kulik wielki*** (*wielki* = 'large', 'great') is in English '**curlew**'.
d) ***orzeł przedni*** – '**prime eagle**' (250) instead of '**golden eagle**'

VII. Kot and Dombrowski (eds) (2001):

a) ***torfowiska wysokie*** (37) – **high peatbogs** (144)
b) ***torfowiska niskie*** (37) – **low peatbogs** (144).

The calque includes the wrong dictionary equivalent.

VIII. Nowiński (2005: 24):

Torfowisko wysokie – **upper peat bog**. Firstly, 'peat bog' is a wrong dictionary equivalent but here we are interested in the translator's further transgression – adding a qualifier 'upper', which makes the whole phrase a calque of the Polish term.

IX. Przystalski and Załuski (1997):

a) ***sikora czarnogłowa*** – '**blackcap tit**' (55) instead of '**willow tit**'
b) ***sikora uboga*** – '**poor tit**' (57) instead of '**marsh tit**'

X. Pudlis (2003: 18):

'Stanowisko **bagna zwyczajnego** w rezerwacie "Czarny Koń"' – 'A **common swamp** in the Black Horse reserve' instead of 'The **Labrador tea** in the Black Horse reserve'.

XI. Sawicki (2003a):

a) zimujące **ptaki wodno-błotne** (80) – winter-abiding **water and mud birds** (81)[63]
b) 'Być może z gęstego **szuwaru** poderwą się kaczki' (88) – 'Maybe some ducks will start from **a rush**' (89). It is difficult to imagine a duck sitting in a specimen of the *Juncus* and *Scirpus* family[64] or

[63] In some sources the translation appears on the page adjacent to the original or a number of pages later in the same book.
[64] They have '-rush' in their names.

even in a thicker *Typha* (bulrush) simply because these plants are too thin in proportion to their height to let a bird bigger than a sparrow perch on its stem, let alone leaves. The plural, rushes, refers to a clump of rush vegetation, which is able to conceal the whole flock of ducks. 'A rush' is a calque of *szuwar*, which both in the singular and plural denotes an uncountable mass of rush vegetation.

XII. Sawicki (2003b):

a) **prace melioracyjne** (66) – **melioration projects** (67). In reality they were **drainage works** and in Polish this is what is commonly understood by the term *melioracja. Prowadzić prace melioracyjne* (76) obtains a similarly misleading equivalent – 'to perform amelioration' (77). The same calque appears in
b) Kłosowscy (2005): 'Oto majowy poranek na Bagnie Bubnów, jednym z ocalonych tu przed **melioracją**, rozległych torfowisk.' – 'A May morning on the marshes called Bagno Bubnów, one of the vast local peat bogs that escaped **melioration**.' (98)

XIII. Sidło, Błaszkowska and Chylarecki (eds) (2004):

a) ***torfowiska niskie*** (57) – **low-lying peatbogs** (59)
b) ***torfowiska wysokie*** (57) – **high peatbogs** (59).

This is a calque combined with the mistaken syntagm/wrong dictionary equivalent.

4.2.1.5. Unnecessary transfers

The following are unjustified borrowings:

I. Durrell (1976: 81) – **hydra –** ***hydra*** (Durrell 1999: 107).

In Polish the creature is called ***stułbia***, which is taught in primary schools. 'Hydra' sounds dangerous and literary, reminiscent of mythological monsters, totally unlike *stułbia.*

II. Durrell (1976: 179):

> 'There were piles of the lovely little golden wild mushrooms, **chanterelles**, with their delicate, almost winy flavour.' – '… drobnych ślicznych złotych

grzybków, ***chanterelles***, o delikatnym winnym smaku.' (Durrell 1999: 179)

Those exotic-sounding *chanterelles* are the commonly collected ***kurki***, officially known in Poland as ***pieprzniki jadalne***. The British, who have no wild mushroom picking tradition, have usually borrowed their names from other languages or adopted them from Latin. In Polish, mushroom names, unless imported for the purposes of Chinese cuisine, are rarely foreign.

III. Fowles (1977) – Fowles (1982):

a) 'In only two days, it seemed, the earth was covered with **anemones**' (60)
   '..., by ziemia pokryła się **anemonami**, ...' (71). Anemones are ***zawilce.***
b) 'a plate of **loquats**' (203) – 'miseczka z **loquatami**' (242) Earlier (83) the translator rendered the same name but referred to trees in Conchis' garden as 'drzewa eryobotrya japonica', which is correct. In order to avoid using the non-existent 'loquaty', she could have written 'owoce drzewa eryobotrya japonica'. However awkward, it would at least have pointed out to an existing plant. But there is a Polish name of the species: **miszpelnik japoński** (also *nieśplik* or *groniweł*). 'Talerz z owocami miszpelnika' sounds quite alien but not so much as the polonised English 'loquaty'. In Durrell's *My family and other animals* (1975a: 222) loquats are ***niespliki*** (1999: 235).
c) 'Everywhere there were flowers – harebells, **gentians**, deep magenta-red alpine geraniums, intense yellow asters, saxifrage.' (258) – 'Pełno kwiatów – **gencjany**, dzwonki, amarantowe alpejskie bodziszki, skalnice.' (308). It is rather odd that while 'geraniums' is correctly rendered as *bodziszki*, though the garden varieties of the flower are called *geranie* in Poland, **gentians** become *gencjany*, although they are named ***goryczki*** and if *gencjany* is used at all, it certainly is not standard nomenclature.

IV. Fowles (1978):

A completely incomprehensible idea is the preservation of English names in the Polish text, especially when the Polish equivalents exist, which is exactly what was done here: 'The thrill for me was finding the rare ones – my first (and last, alas) Monkeys near Goring, a solitary Fly under a sunshot whitebeam at the edge of a Chiltern beechwood. His heaven was a wet meadow full of dull old Dactylorchids: counting and measuring and noting down the degree of hybridization.' (71) – 'Moją pasją było znajdowanie rarytasów: jak na przykład pierwsze (i ostatnie, niestety) ***Monkeys*** nieopodal Goring czy samotna ***Fly*** pod osłonecznioną mąkinią na skraju buczyny w Chiltern.' (Fowles 2000: 89). In 'dull old Dactylorchids' **Dactylorchids** remains untouched in the Polish text. The exotic-sounding Monkeys (monkey orchids) are ***storczyki małpie***, Fly orchid is ***dwulistnik muszy,*** while Dactylorchids are ***kukułki***. Since the last name is quite unfortunate – *kukułki* are primarily very well-known birds – the Latin *Dactylorhiza* could be used. Although English names of plant varieties obtained by British growers often penetrate into other languages unchanged or adapted, names of species must be quoted either in the native tongue or in Latin according to the international system.

V. Fowles (1978):

> They had gone down to Hampshire in pursuit of the elusive **Summer Lady's Tresses**, one of the rarest British orchids. A long and azure weekend, interminable scrambling through bogs and over grass-tussocks; not a sight of the plant. (181)

> W bezskutecznym poszukiwaniu ***Summer Lady's Tresses,*** jednej z najrzadszych brytyjskich orchidei, dotarli do Hampshire. Długi, lazurowy weekend, nie kończące się brnięcie przez kałuże oraz kępy traw – i ani jednego kwiatu (Fowles 2000: 219).

The orchid which Daniel and Anthony[65] were searching for is in Polish ***kręczynka letnia,*** *Spiranthes aestivalis,* a close relative of the mysterious *Spiranthes spiralis* mentioned on page 9. The borrowing is totally unjustified. Incidentally, the plant is *storczyk* not *orchidea* and the hobbyists were walking through the bogs not just puddles, as the translator imagines. Earlier, on page 69, Daniel entered Anthony's room at Oxford and saw 'a stem of the Man Orchid, Aceras' in a jam-jar. In the Polish version (2000: 87) it becomes 'gałązka orchidei Aceras'. In this case the Polish name does not exist, so one ought to resort to Latin and write 'łodyżka (not gałązka = a twig) storczyka zwanego *Aceras anthropophorum'*. In order to avoid identification problems *Aceras* will

[65] Daniel is the main character of *Daniel Martin*, whereas Anthony is his friend, an academic teacher of philosophy.

not suffice because, for instance, in Blamey, Fitter, and Fitter (2003) the orchid's Latin name is given as ***Orchis** anthropophorum.*

VI. Fowles (1978: 181):

Daniel answering Anthony's question concerning his country property and whether it is good orchid country says: 'Greenwings in one meadow. One quite nice colony of *spiralis*. When the sheep don't get them. Early purples.' Greenwings are green-winged orchids – ***storczyki samcze***, *Spiranthes spiralis* (common lady's tresses), as discussed in 4.2.4.1.2. (p. 246), is ***kręczynka jesienna***, while early purple orchids are ***storczyki męskie***. The Polish translation is as follows: 'Jest łąka z *greenwings*, a także całkiem ładna kolonia *spiralis*, jeśli oszczędzą je owce. Są *early purples*.' (Fowles 2000: 219). While native speakers of English – hobbyists and specialists – immediately recognise the plants discussed, Polish readers enjoying the same status, remain clueless (except in the case of *spiralis*), until they consult English sources.

The translator has naively adopted the techinque employed relatively successfully in translating the cultivars of fruits such as plums (Fowles 1978: 4 – Fowles 2000: 11) or apples and pears (83) – (103). Cultivars are often confined to a particular region or small locality and they simply have no equivalent in foreign languages.

VII. Fowles (1996):

> The ground about him was studded gold and pale yellow with **celandines** and **primroses** and banked by the bridal white of densely blossoming **sloe**; where jubilantly green-tipped **elders** shaded the mossy banks of the little brook he had drunk from were clusters of **moschatel** and **woodsorrel**, most delicate of English spring flowers. Higher up the slope he saw the white heads of **anemones**, and beyond them deep green drifts of **bluebell** leaves. (71-72)

> Ziemia wokół niego usiana była złotymi i bladożółtymi cętkami **chelidonii** i pierwiosnków, obramowana zaś panieńską bielą tarniny w pełnym rozkwicie. Tam, gdzie krzewy czarnego bzu o lśniąco zielonych czubkach ocieniały brzegi strumyczka, z którego zaczerpnął wody, rosły kępy piżmaczka i szczawika, najdelikatniejszych spośród angielskich kwiatów wiosennych. Wyżej na zboczu dostrzegł białe główki **anemonów**, a za nimi ciemnozielone pasy liści dzwonków (Fowles 2003: 65-66).

Most of the plants bear correct Polish names except the two cases printed in bold, where quite plain names of common flowers are replaced with

Latin-based sophisticated-looking ones. The effect is almost comic. **Celandines** are simply ***ziarnopłony wiosenne***, common heralds of spring, **anemones** – ***zawilce*** (this mistranslation appears again on page 68).

4.2.2. Misinterpretation errors

This category contrasts with the previous one in that the latter stems from the lack of interpretation. The causes of both remain essentially the same – insufficient knowledge and bad intuition concerning the target culture.

4.2.2.1. Mistaking two source language syntagms or verb frames

I. Chylarecki and Sawicki (2003):

a) 'gęstwina lasów łęgowych z **podmokłościami** nadającymi się doskonale do błotnych kąpieli' (38) – 'The thickets of riparian forests with their **swamps** ideal for mud baths' (39).
Swamps are not just wet sites, which is what the Polish word printed in bold conveys. The translator should have discriminated between *podmokłość* and *bagno* despite fairly fuzzy borderlines separating the two categories.
b) 'Od poprzedniego gatunku różni ją to, iż jest **smuklejsza** (choć 30-procentowa różnica w wadze nie przekłada się na równie dużą różnicę wielkości ciała).' (54) – 'Compared with the former [the ringed plover] they [little ringed plovers] are a **bit smaller** (however, their 30% **more** weight is not adequately expressed by their bodily proportions)' (55). The English reader obtains completely disparate and incorrect information. Firstly, to be precise, the little ringed plover (*sieweczka rzeczna*) is 'more slender' rather than 'a bit smaller'. Secondly, and more importantly, it does not weigh 30 per cent more than the ringed plover (*sieweczka obrożna*) but less. This difference in weight is not easily noticed since its body is not proportionally smaller. The translator confused the referents and misinterpreted the syntagm '30-procentowa różnica'.
d) 'W rezultacie coraz większy odsetek zniesień rybitw, sieweczek, mew i innych siewkowców ...' (77) – 'In effect the ever-increasing number of eggs of terns, shore birds, gulls and plovers ...' (75-76). In the original the message is straightforward: the problem concerns the eggs of terns, plovers, gulls and other shorebirds. In translation, plovers, terns and gulls form a category of their own

rather than belonging, as they do, to the shorebirds (*Charadriiformes*). The taxonomy of birds, as of any other living organisms, changes constantly together with the progress in scientific research. That in itself causes difficulty in grasping the correspondences between the names of various groups of animals in the two languages, let alone mistakes like the one quoted. ***Siewkowce*** is a suborder of ***siewkowe*** (the order of *Charadriiformes*). In English they have the same name: **waders** or **shorebirds** (preferred in North America). The other suborders are ***mewowce*** **(gulls)** and ***alki*** (**auks**). One of the families of ***siewkowce*** is ***siewkowate*** (*Charadriidae*), which in English functions as **plovers** and is divided into three subfamilies (*Charadriinae*): lapwings, plovers (***sieweczki***) and dotterels. Where in Polish we have four different words, in English only two appear, and one of them comes with a synonym. For the sake of clarity the two taxonomies and their interrelations may be represented as follows:

siewkowe => *siewkowce* => *siewkowate* => *sieweczki*

waders/shorebirds waders/shorebirds => plovers => plovers

In Sidło, Błaszkowska and Chylarecki (eds) (2004) **siewki** (listed as follows: lapwing, black-tailed godwit, barred-tailed godwit, redshank) and ***ptaki siewkowe*** are both translated as **waders** (163), later (378, 430) **.**

II. Durrell (1974) – Durrell (2002):

'The avenue of **bamboo**' (115) – '**trzcinowej** alei' (76)
'squeaking, creaking, groaning and rustling avenue of **bamboo**' (116) – 'poskrzypujących, trzeszczących, wzdychających i szemrzących **trzcin**' (77)
'a **bamboo** fence' (117) – '[działce] ogrodzonej **trzcinami**' (78)
'a **bamboo** gate' (118) – '**trzcinowa** furtka' (78)
'I looked up and saw giant **bamboos** that curved over the road' (155) – '… gigantyczne **bambusy** ...' (104)
'a tiny **bamboo** cage [with a pigmy owl inside]' (177) – '… **bambusowej** klatki ...' (118)

'Bamboo' is in Polish ***bambus***. ***Trzcina*** is 'reed'. There is also a plant called ***trzcina bambusowa*** – bamboo reed. Durrell is consistent in using 'bamboo' in all the contexts quoted, whereas in the Polish version we have mostly *trzcina* and twice *bambus*. Although the two plants are similar, the bamboo is usually much more woody and durable and thus more likely to be used for the construction of fences and gates than reed. By rendering 'bamboo' as *trzcina* the translator has introduced an intriguing extra element but probably detracted from the truth.

III. Durrell (1975a) – Durrell (1991):

Giant snails (59) – ***gigantyczne węże*** (86) (**giant snakes**). The error is later repeated: 103 – 153 the hairy frogs eat snails – '… *jedzą węże*'
The difference between the two language versions in both cases is shocking. However similar the words 'snails' and 'snakes' look and sound, mistaking one for another twice smacks of mockery.

IV. Durrell (1975b) – Durrell (1999):

a) 'of their [goats'] mouths ripping and tearing at the **undergrowth**' (45) – '… szarpania i gryzienia **trawy**' (45)
   Here, it must be something more difficult to graze than plain grass, which is signalled by the word 'undergrowth' – ferns, shrubs and saplings of trees.

V. Fowles (1975: 237) – Fowles (1994: 255):

The children have found a snake by the river. Sally, Peter's girlfriend asks Annabel, the children's mother: 'Aren't they [snakes] dangerous?' and is given the following answer: 'They are only grass snakes'. That implies that the snake the children are looking at will not poison them. Annabel must have known, uttering the words, that away from the river among the dry rocks poisonous vipers live, as Peter discovered during his walk there later in the book (page 279). In the Polish translation Sally asks: 'Nie ukąsi ich?' and Annabel responds: '**Tu** są tylko zaskrońce' (= there are only grass snakes here), instead of '**To** są tylko zaskrońce'. ***Tu*** may mean both by the river and away from the river but in the surroundings. In English, the referent is clearly defined, while in Polish, contrary to what we know about Annabel, she appears to be either ignorant or deceptive or absurdly reassuring, verging on recklessness. Misread syntagms have consequences for the reader's perception of one of the main characters.

VI. Fowles (1975) – Fowles (1994):

a) 'The sparkling water, the splashing feet, the dragonflies and butterflies; the **buttercups** and oxeyes and little blue flowers like splashes of sky.' (245) – 'Musująca woda, chlupoczące stopy; ważki i motyle; **orchidee**, złocienie i niebieskie kwiatuszki jak odpryski nieba' (264).
The inappropriateness of the word *orchidea* to denote an orchid has already been discussed but using it as an equivalent of 'buttercup' is a serious error. The buttercup is ***jaskier*** and does not even distantly resemble any of the European orchids.
b) By making the following comment (Fowles 1975: 235): '[An oriole] gave its particular combination of heat, water and foliage a voice, defined exactly its foreignness, its faint subtropicality – so leafy, so liquid, so richly of its place and season, Central France and late May' the author has established the significance of the oriole's voice in the book. In spite of that the translator treats the rendering of it in Polish lightly:

> ' "Listen", says Bel, "there's an oriole" ... They hear the **liquid whistle** from across the river' (254). – ... Słyszą **potoczysty trel** zza rzeki' (Fowles 1994: 273).

The oriole's song is indeed **'liquid whistle'**. Far from *potoczysty trel* (rich fluent trill), it can be in Polish ***melodyjny/fletowy gwizd***. The message from the narrator is that subtropics are suffused with an exuberance of whistles more than trills. In Polish, no connection is established between the liquidity of the oriole's voice and the perception of Central France as subtropics. For the translator the two distinct syntagms: liquid whistles and rich fluent trills apparently amount to the same.

VII. Fowles (1977) – Fowles (1982):

> 'The **antlers** [of a stag] were enormous, ... , many-tined' (499) – **'Rogi** były ogromne ... , rozgałęzione' (138 (vol. 2)).

'Antlers' is ***poroże***, ***rogi*** – horns. Antlers are filled inside, horns are empty when taken off their base. It is true that there is no fear that people would evoke the image of a cow's horns on the head of a stag. But promoting wrong terminology in the book written by the author so sensitive to natural diversity and minute differences within the natural world is quite irresponsible. We cannot be sure whether antlers were mistaken for horns or *poroże* for *rogi*. In the latter case the error would have to be moved to the class of target language realisation errors.

VIII. Fowles (1978) – Fowles (2000):

a) **roebuck** (286) – ***jeleń*** (289) instead of ***kozioł*** (*sarny*). The roebuck could have been mistaken for a stag or the error might have crept in at the level of target language syntagms (if *jeleń* had been thought to stand for the male of the roe deer).
b) The following is about Anthony, already deceased: 'But he had a much more genuine hatred, almost a fear, of what could not be collected, classed, precisely defined, noted down: I mentioned earlier his obsession with the Dactylorchis group of British wild orchids, which hybridize with bewildering frequency – I think the fluid frontiers between their **species** seriously upset him, and they were rather less a botanical challenge than a nagging flaw in his would-be highly ordered nature of things' (361). In Polish, we read 'płynne granice między **odmianami**' (433). The whole point is that the frontiers are so fluid between the different species, not just varieties. This illustrates the fuzziness of categories such as species, subspecies, kinds, varieties. Encouraged by such flexibility, the translator probably concluded that since hybridisation makes the frontiers between the species fluid it did not matter whether we called something a species or a variety – after all, what was at a given moment professed a variety might subsequently be recognised as a species and vice-versa. In my opinion, such aggressive relativism must be curbed and until a species remains a species, in Polish it stays *gatunek* (not *odmiana*), too.

IX. Iwaszkiewicz (1969) – Iwaszkiewicz (2002):

a) 'zapach kwiatów **bobu** i koniczyn' (152) – 'the scent of **bean** flowers and clover' (179)
   Bób is **broad bean,** while **bean** is *fasola*, not necessarily building the same image.
b) 'stał teraz w oknie hotelu, pachnącego waniliową wonią **świerkowych** desek' (313) – 'Now he was standing at the window of the hotel, redolent with the vanilla odour of **pine** wood' (240). To Iwaszkiewicz it was the **spruce** wood that smelt of vanilla, not the pine.

X. In Kłosowscy (1999: 128):

a) a photograph shows a sparrowhawk with its young. They are still covered with down but the caption goes: 'A sparrow-hawk with its fledgling'. Firstly, there are more than one and, secondly, by definition fledglings already have feathers that enable them to leave the nest. In order to produce the correct version the translator should have been more faithful to the original. 'Krogulec z młodymi' is simply 'A sparrowhawk with its young'. In this way confusing **nestlings** (***pisklęta***) with **fledglings** (***podloty***) is avoidable. The similarly flawed translation of *pisklęta* appears in the summary of: Zawadzcy (2003: 143) – 'Fledglings [of the capercaillie] hatch out after 25-26 days. Supervised by their mother, a family flock soon ramble the area in search of food. The fledglings develop very quickly'. Although precocial birds, capercaillie chicks emerge from their egg shells as downy creatures with no flight feathers. The use of the word *podloty* in this context would raise a few eyebrows among Polish naturalists. The same effect on English bird lovers can be expected with fledglings instead of chicks.

b) Kłosowscy (1999: 250):
As an illustration that life finds its way everywhere the authors give 'kobierzec mchów wątrobowców i **krzewinek** wpełzających na wielkie głazy Karkonoszy.' One can expect '**under shrubs**' or '**dwarf shrubs**' on such boulders and not **brushwood**, as the translation has it – 'the carpet of liverwort and brushwood spread over large boulders in Karkonosze.' 'Brushwood' suggests 'a thicket composed of shrubs and small trees' (Webster's), so something much taller and more impenetrable than a type of vegetation that can settle well on a boulder. *Krzewinki* and *krzewy* are not the same things, especially when one envisages them in larger concentrations.

XI. Kłosowscy (2003: 141):

'Zdecydowanie częstszy jest on [podróżniczek] **na wschód i północ od naszych granic**, ale dolina Biebrzy jest miejscem jego dość licznego gniazdowania' is rendered as: 'Although it [the bluethroat] is definitely more widespread **in the north and east of Poland**, the Biebrza valley is a rather popular nesting site for it', which misleadingly suggests that the Biebrza is not located in north-eastern Poland. The translator has overlooked the fragment that the bird is more widespread to the north and east 'od naszych granic' – 'of our border' or, in other words, misread it as 'within our borders'.

XII. Kossak (2001):

a) 'Po nim puszczają szkliste okowy w **borach** i **grądach** ...' – 'Then the icy bonds were cast off in **woodlands** and **wet-ground forests** ...' (28) *Bory* are needleleaf stands and translating them as woodlands is not very precise, but when they are contrasted with *grądy*, which are essentially broadleaf stands, such a translation is most undesirable. The point is, however, that the contrast broadleaf/needleleaf is lost in translation. Instead, we are puzzled by a false contrasting of woodlands and wet-ground forests. After all, wet-ground forests are woodlands, just as dry-ground ones. *Grądy* and *bory* can both be wet and dry. The clause that follows makes the woodland/wet-ground forest even more peculiar: '..., while on peatbogs and riverine forests ice would sometimes linger till May.' Riverine forests are definitely wet-ground. Thus, is ice cast off earlier or later in wet-ground forests? Certainly, the readers of the Polish original do not ask such a question. They understand that in drier *bory* and *grądy* ice disappears earlier, whereas in wet *łęgi* (riverine forests) significantly later. The translator does not notice the unsuitability of the term 'wet-ground forest' to denote *grąd* even when, on the same page, she writes that in it: 'groundwater stays rather deeply [and only] at times it is stored in the root zone due to poorly transmissive clays.' A similarly wrong image is created by the following mistranslations in Sidło, Błaszkowska and Chylarecki (eds) (2004: 237) ***bory i grądy*** – **coniferous and dry-ground forests** (239). The misread syntagm *grądy* leads to the false contrasting of coniferous forests and dry-ground forests (in reality oak-hornbeam forests), as a result of which, to the reader's astonishment, coniferous forests appear to be inherently wet.
grądy (241) – dry forests (243)
grądy subkontynentalne (245) – dry subcontinental oak stands (248). *Grąd* is not just oak forest
'lasy grądowe z dominacją sosny i udziałem dębu, grabu i lipy' (272) – dry forest predominated by ... (274)
grądy (423) – oak-hornbeam forests (425); oak stands (518)
lasy grądowe (520) – dry oak forest stands (522)

zbiorowiska grądowe (546) – dry oak stands (547)

b) '... started to wade the water in the **rushes**' – '... wskroś **trzcin**'. It should be '**reeds**' (136)

c) On pages 201 and 386 '**bucks**' used in reference to red deer males degrades them to roe deer; earlier (199) '**harts**' was used in the sense *byki.* On page 240 it is even more explicitly wrong: 'fighting **bucks** before timid hinds'. Hinds are red deer females.

d) On pages 298, 369, 386, 493, 540 a **doe** is *łania* (actually, this is a female roe deer);

e) female **roe deer** are inappropriately called **hinds** (339);

f) male roe deer are called **stags** (340) – it seems the translator has problems with coming to terms with the differentiation buck/stag and doe/hind irrespective of the language, which the correct translation of ***kozły*** as ***roebucks*** only four pages later seems to corroborate.

g) ***jelenie*** are rendered as **deer** (343), which is the umbrella term for many species called 'deer';

h) Let us consider '**wyłaźbił barcie**' and '**za udział w patoce**' in 'Last autumn his [Ławrysz's] neighbour **cleaned the bee nests** and protected them from the cold with bundles of dry heather in exchange for **his help in honey straining**' (223). The fragments of the sentence printed in bold in the English version are easy to comprehend. The Polish reader must wait for explanation until later in the book or consult a good dictionary. When s/he reaches page 228 s/he will be rewarded with the information that *łaźbić* means *podbierać* (take some of the honey combs) and that *patoka* is strained honey. As we can now see, in English the message is different. The neighbour 'cleaned the bee nests' rather than collected honey combs and helped Ławrysz in exchange for part of the strained honey, not because Ławrysz had helped him in straining. The standardisation of language is sometimes desirable, but if it leads to falsification through changes in imagery, it cannot be justified.

j) 'The best ever **July** honey comes from ...' (232) Honey is not normally labelled on the basis of the harvest time. Where the foreign reader may think a local peculiarity is presented, the original tells us about *miód* ***lipowy***, i.e. lime honey not *miód lipcowy*, nothing culture-specific. The translator must have confused two adjectives, whose spelling differs by just one letter but whose meaning by a great deal.

k) Spruces are said to form admixtures in '**fens** and **marshy meadows**', which is supposed to stand for '**łęgi** i **olsy**' (326). The images imposed by 'fens and marshy meadows' are of rather open wetland, not woodland. Fens are a type of peatland. The term *łęg* is ambiguous and can mean 'marshy meadow', but common sense, which must have left the translator at this point, tells us that it is not this meaning we ought to look for. *Łęgi*, just like on a number of pages before, should be 'riparian/riverside forests, *olsy*, in turn, 'alder swamp forests'.
l) 'na 1 kozła przypadało 6-7 kóz' – 'the ratio of males to females [of roe deer] was 6 to 1' (452). It is the other way round. The order of nouns remains the same in both languages despite the introduction of the new subject in English, which may have put the translator on the wrong track.
n) Conventionalised imagery is reflected in grammar. Wrongly interpreted in a foreign language, grammar can distort images and, in consequence, messages conveyed. Here is an example. On page 476 there is a description of how one of the wardens (Jan) managed to prevent a deputy prime minister in the 1970s, Kazimierz Secomski, from killing the last of the capercaillies surviving in the Białowieża Forest. In the Polish version we have: 'Już towarzysz premier podnosił broń do oka, już widział kolejne trofeum ...'. *Już* plus past imperfective signals that something was just about to happen, but it did not, due to some interruption. In translation we read:

> Every time comrade minister raised his gun to aim at something, imagining another trophy on his study wall, when Jan, struggling with his enormous hangover, shouted histerically:
> 'Fuck you! This is the last capercaillie in the whole of the Białowieża Forest! Shoot, son of a bitch! I'm going to booze'.

First of all, 'when' is grammatically incorrect here. More importantly, the use of 'every time' (why not 'each time'?) and 'at something' suggests that there were numerous attempts to raise the gun to aim at various targets and that the scene following them repeated itself many times. Obviously, the same capercaillie would not have let intruders bother him even one more time. Once would have been enough to lose sight of the bird and any chance of shooting it. Incidentally, the bird has excellent hearing and to scare it away a whisper suffices. The grammatical error – misread syntagms – destroys the original image causing total misunderstanding on the part of the target reader.

XIII. Miłosz (1989) – Miłosz (1981):

a) '... szczelnie obrosłe **łoziną** ...' (7) – 'The Issa is a deep, black river with a lazy current, thickly bordered with **reeds**;' (5)
Łoza is a species of willow, *łozina* being its higher concentration. River banks overgrown with willow scrub look entirely different from those constituting reed beds. The average reader, the translator included, tend to perceive *łozina* vaguely as an impenetrable thicket, no matter its composition.

b) 'Raz na kupie kamieni na pasiece znalazł gniazdo dudków – sięgnął ręką i złapał **małego, co nie umiał jeszcze latać**, rozkładał tylko wachlarz na głowie, żeby straszyć.' (36) – 'Once among a pile of rocks in the brush, he discovered a nest of hoopoes; reaching inside, he caught one (hoopoe) in his hand, a **fledgling**, which tried to frighten him by fanning its head feathers.' (36)
In the Polish text there is **'a young, which was not able to fly yet'**, which in English becomes 'a fledgling'. The English word means 'a young, already feathered bird that has left its nest'. The translator must have thought that if the hoopoe had been able to fan its head feathers, it sufficed to interpret it as *podlot* ('fledgling'). The fact that it had not left its nest and, by implication, had not been ready to fly, did not persuade him that it still belonged to the category of *pisklęta* ('nestlings') or, simply 'the young', a conveniently underspecified term.

c) 'zaciosany koł z pnia **osiny**' (57) – 'a spike carved from an **alder**' (61)
This object was used to pierce the dug-up body of Magdalena, who had committed suicide, because she was believed to haunt and scare people. Using this method against vampires is well known in some cultures. The point here is that the translator changes the species of the tree the spike is made from and by employing the word 'spike' (thin) instead of 'stake' (thick) evokes a different image in people's mind. One might presume that such rendering is imposed by cultural differences, but the inspection of the relevant sources[66] does not confirm this presumption. Through the vampire's heart an aspen (*osina*), ash (*jesion*) or white thorn (*acacia constricta*) stake are supposed to be driven. It is also possible to make the stake from hawthorn (*głóg*), maple (*klon*), blackthorn (*tarnina*) or buckthorn (*szakłak*). Alder (*olsza*), is not

[66] See e.g. http://www.madametulbot.com/pix/posters/vampire2.htm_11k

mentioned at all, whereas aspen is given prominence. Interestingly, the translator knows that ***osina*** (158) is aspen (174).

d) 'mały dworek z drewnianymi kolumienkami ganku, obrośnięty **czarnym bzem**' (115) – 'The small manor house was fronted by wooden porch columns overgrown with **purple lilacs**' (126)
Czarny bez is 'elder' and looks very different from purple lilacs – the elder (the honeysuckle family) has inconspicuous white-yellowish flowers, light green pinnate leaves, black berries as fruit used to produce medicinal juices, the purple lilac (the olive family) has showy, purple blooms, dark-green ovate leaves and unattractive inedible fruits. Both include *bez* in their names, which must have caused the misinterpretation.

e) ***rządowy las*** (125) – **the preserve** (138)
The word used implies that the forest was a nature reserve, where natural resources were preserved for future generations, whereas, in fact, the forest was managed (exploited, not just preserved) by the state and was not owned privately. In a sense, its timber stock was preserved for the exclusive use of the government and that may have instigated the choice of the word. The failure to spot ambiguity may point out to it being a realisation error, which best illustrates the lack of sharp division lines between the categories in the analysed taxonomy.

f) 'Te same wykroty i plamy mazi pod skłębionymi pękami **łóz**' (160) – 'The tree craters and swampy patches congested with **rushes**' (176). *Łozy* are grey willows not rushes (see XIIIa) above for comment).

g) 'wysokie **trzciny**' (162) – 'tall **rushes**' (wrong) (179)
'uciec w **trzciny**' (162) – 'taking cover in the **reeds**' (correct) (179)
'Już odróżniało się cyple **trzcin**' (179) – 'As soon as they made out the gaunt shapes of
bulrushes'(wrong) (197)
'zagradzała drogę ucieczki do **trzcin**' (179) – 'cutting off their escape route into the **rushes**'(wrong) (197)

h) 'zakamarków w **sitowiu**' (163) – 'sanctuaries among the **reeds**' (180) (wrong – properly: club-rushes, 'rushes' acceptable). In Fowles (1978) – Fowles (2000) the situation is reversed '[Mark] climbs down into the reeds' (24) becomes 'schodzi w sitowie' (33).

i) 'moorland' and 'moor' introduced into the English version, following the misinterpretation of a few Polish syntagms, bring about a muddle:

> Być może kiedyś było tutaj jezioro, ale na jego dawnym dnie teraz rozpościerały się łąki z ostrą wikswą i dalej, przed nimi, **mszaryna** – kępy, na których rosły karłowate sosenki, tu i ówdzie chaszcze splątanej **łozy**. ... Jest to królestwo zapachu. Z mchu sterczą tu krzaki bahunu – *Ledum palustre* – z ich wąskimi skórzanymi liśćmi, niebieskie jagody pijanic, duże jak gołębie jajka, dojrzewają w cieple powietrza przesyconego parą. Młode cietrzewie znajdują tu dość jedzenia, pod kierownictwem matki ... (214)

> Once a lake, its former basin was now a sprawling meadow covered with sharp sedge and, up ahead, with moss, **thick with pine scrub** and clumps of tangled withes. ... The **moorland** was a kingdom of fragrance. The moss sprouted bushes of marsh tea (*Ledum palustre*) with narrow leathery leaves, and blue bog bilberries the size of pigeon eggs thrived on the warm, vaporous air. ... The **moor** made a good feeding ground for the young **blackcocks**, still under their mother's guardianship ... (235)

Clearly, the above is a description of **bog woodland** bordering on the raised bog and not of moorland. Miłosz uses the word ***mszaryna,*** which may not be understood fully by all Polish people but they would at least imagine something vaguely resembling what scientists call bog woodland. Foreign readers of the translation will imagine something that may not be even vaguely similar, especially as the translator has chosen to replace '(mossy) hummocks overgrown with dwarf pines' with 'thick with pine scrub'. A moor thick with pine scrub ('wyżynne wrzosowisko gęsto porośnięte karłowatą sosną') is a contradiction in terms.

Another puzzle is the presentation of the habitat as the feeding ground for young **blackcocks**, as if young females of black grouse fed somewhere else without their mother's guardianship. The plural form *cietrzewie* can denote both sexes as a species or the males only, depending on the context, which, here, points out unequivocally to the first option. There is also a problem with the name ***łoza***, this time rendered as mysterious 'withes', which is absent from Anioł-Kwiatkowska (2003), Blamey, Fitter, and Fitter (2003) and Moor (1984), and means 'a flexible slender twig as that of the osier' (Webster's). As to ***mszaryna***, when it appears in the context of hunting snipe – it is rendered satisfactorily as '**moss**' – 'szli coraz dalej w mszarynę' (215) – 'they kept plowing through the moss' (236).

XIV. Narodowa Fundacja Ochrony Środowiska. *Biebrzański Park Narodowy* (1996): ***Śródbagienne*** *grądy* (1) – oak-hornbeam forest **on bog** (12)

First of all, *śródbagienne* means '**occurring among swamps or marshes**', not in them. Secondly, oak-hornbeam forests like nutrient-rich, heavy, clayey soils and would never grow on bogs, by definition nutrient-poor habitats. Thus, the correct version should be: 'oak-hornbeam forests among swamps and marshes'. This error could equally well be classed as insufficient knowledge of the subject-matter.

XV. Ośrodek Edukacyjno-Muzealny RPN (2000) (a map):
Tereny podmokłe – **swamps**. These are **wetlands**, swamps being an exaggeration.

XVI. Pudlis (2003):

a) We are informed that '[t]he first stopping place is a culvert on a small stream' (27). It is puzzling since culverts rather than retaining water enable it to escape. The Polish reader is not confused because s/he reads: 'Pierwszy przystanek to **zastawka** na niewielkim strumieniu', which matches the context – the rising awareness and intensified endeavours of foresters to retain water in woodlands. *Zastawka* means 'weir', a structure used to store water in a small reservoir by arresting its flow. '**Culvert**' in Polish is '**przepust**'.
b) An educational trail presents 'the basic rules for management and protection of **coniferous stands in woodland habitats**' – **'drzewostanów iglastych na siedliskach borowych'** (22)
'W LKP dominują **siedliska borowe** ... ; pozostała część to **siedliska lasowe** ze znacznym udziałem **borów**, lasów wilgotnych mieszanych i wilgotnych bagiennych.'

As in b), in translation the fact that needleleaf stands can occupy habitats naturally taken by broadleaved trees is ignored: 'This PFC is dominated by **coniferous habitats** … with the remaining areas being **broadleaved forest habitats** with a considerable proportion of mixed and moist broadleaved forests.' (33)

The folk understanding of the word *bór* influences several other translations to the extent that translators forget that they are dealing with specialist contexts:

c) Kossak (2001: 28):
'Po nim puszczają szkliste okowy w **borach** i grądach ...' – 'Then the icy bonds were cast off in **woodlands** and wet-ground forests ...'
d) Sobociński (2004: 7):
'W różnych typach **borów** dominującymi drzewami są ...' – In various types of **woods** ...
e) Sawicki (2003b):
'żywiczny zapach **borów**' (84) – 'resinous smell of **forests**' (85). The word 'resinous' should have opened the translator's eyes to what *bory,* contrary to what some people believe, actually mean. They are **stands of conifers**.

XVII. Sawicki (2003a):

a) 'w bezodpływowym zagłębieniu na południe od jeziora Mikoszewskiego leży otoczone **podmokłościami** niewielkie oczko wodne.' (24) – 'another small pond surrounded by **swamps** exists in a depression without outflow to the south of Jezioro Mikoszewskie.' (25). Although lakes are often surrounded by swamps, here the image drawn by the author is that of an area that is just a little wet. The translator overinterpreted the word *podmokłości* or underestimated the degree of waterlogging of swamps – again the distinction between misinterpetation and realisation errors is blurred (see also I. above). Moreover, the depression has no outflow at all, not just to the south of the lake, as the rather lame syntax of the translation implies.
b) 'tatrzańskie **hale'** (54) – 'the Tatra **coombs**' (55). ***Hale*** are mountain meadows and pastures and can be found in coombs (deep valleys or basins) as well as on mountain slopes. The translator must have confined the mountain meadows only to valleys, where undoubtedly most of them spread. In parallel, he could have ascribed a different meaning to the target language word 'coomb'.
c) On pages 64 and 65 the term ***lęg*** causes the translator a great deal of trouble, probably because he confuses three Polish syntagms: *wylęganie się* ('hatching'), *lęg* ('brood') and *lęgi* as in *wyprowadzać lęgi* ('to breed successfully'):
   - 'dla ptaków wyprowadzających tu lęgi' he renders as 'for the birds which **breed and hatch** there'. Hatching is a necessary part of breeding, so the latter is sufficient.

- 'porzucenie **lęgu**' – 'abandonment of **the eggs'** – here it is correct
- 'nie mogą osiągnąć **sukcesu lęgowego**' – 'hatching failures' – ***Sukces lęgowy*** involves much more than hatching. A bird breeds successfully if it manages to rear its young until they fledge and/or start living by themselves. The translator constantly confuses hatching for breeding, also later, where *sezon lęgowy* (92) becomes **hatch** season (93).

XVIII. Sawicki (2003b: 24 and 25) inadvertently promotes a peculiar understanding of the modifier *niskie* in ***torfowiska niskie*** translating it as '**lowland peat-bogs**'. The same problem occurs in Sidło, Błaszkowska and Chylarecki (eds) (2004: 239): *torfowiska niskie* – lowland peatbogs. It must be noted that raised bogs could also be lowland and here it is unlikely that the reader of the English text would guess the actual referent – fens. Kucharski and Pisarek (undated) make the error even more serious by calling *torfowiska niskie* (7, 23) – lowland bogs (27, 28) and so does Narodowa Fundacja Ochrony Środowiska. Biebrzański Park Narodowy (1996): *torfowisko niskie* (1, 9) – low bog (12, 20) or *torfowisko niskie* (2) – low peatland (13). Ośrodek Edukacyjno-Muzealny RPN 2000 (a map) beats all the others by translating *torfowisko niskie* as lowmoor and *torfowiska wysokie* as thighmoor (a typing error to cap it all?).

XIX. Sawicki (2003b):

a) 'buczący samiec bąka' (38) – '**humming** male' (39)
b) 'odzywające się samce derkacza' (42) – '**humming** males' (43). The sound produced by the bittern (*bąk*) is '**booming**', the voice of the corncrake (*derkacz*) is **loud, harsh and grating** – definitely unlike the gentle humming of insects in the trees. The translator was either unaware of the sounds uttered by the birds and misinterpreted their verbal notation or failed to match the Polish words with their optimal English linguistic representation.
c) '**tokujące** samce derkacza' (42) – '**tooting** males' (43). The word tooting is used in reference to male birds which perform courtship displays including vocal contests. However, it seems to be restricted to certain American species. The voice uttered by the corncrake does not resemble tooting, anyway.

XX. Sidło, Błaszkowska and Chylarecki (eds) (2004):

a) **łęgi olszowo-jesionowe i olsy** (241) – **alder-hornbeam and alder carrs** (243)
Hornbeam (*grab*) is a different species from ash (*jesion*) and does not co-occur with alder, while using 'carr' in reference to both types of woodland – *łęgi* and *olsy* – breeds confusion. Likewise, on page 308 ***łęgi*** is rendered as **alder forest.** In Polish, we receive very concrete information, while in English the message is general and might just as well bring to mind any patch of wood consisting of alder trees. On page 557 there are ***łęgi wierzbowo-topolowe***, but on page 558 the English version informs us about 'willow, **alder** riverine forests'. The alder imposes a completely different look on the place than the poplar (***topola).*** It is difficult to determine whether the translator constantly confuses species of trees in his native tongue or does not know them in the target language.

b) ***lasy mieszane*** (359) – mixed **deciduous** forest (361). Why did the translator come to the conclusion that these forests were only deciduous if the very nature of mixed forests is that they include both broadleaved and needleleaf trees?

XXI. Sobociński (2004):

a) 'To wspaniały, obfitujący w zespoły lasów **mieszanych**, zróżnicowany siedliskowo **relikt** pierwotnych puszcz.' (4) – 'This is a magnificent **artifact** of primeval forests, with varied habitats, rich in **composite** woods' (5). Primeval forests are not artifacts but natural systems. The Białowieża Forest is a relic primeval forest. Although I have located this error in the class of misinterpretation errors, it may have been made at the stage of looking for an equivalent. Returning to the Białowieża Forest, it is composed of **mixed forests,** that is both broadleaved and needleleaf ones. 'Composite woods' is not a term but a rather unconventional and ambivalent phrasing.

b) '... białowieskie grądy, olsy, łęgi, bory, a także nadrzeczne łąki i bagna naprawdę są **ostatnią** puszczą.' (6). – '... the alder and oak woods, riverside meadows and swamps really **make the ultimate** forest' (17). 'Ultimate' means ***ostateczny,*** i.e. final, eventual, incapable of further division or development. ***Ostatni*** means 'the last' here: 'the last of the kind' but no doubt undergoing constant transformational processes according to the laws of nature and due to human impact.

c) 'Puszczańskie wilki co roku eliminowały 10% **jeleni** ...' (6) – '[in the 1990s] the wolves eliminated 10% of the **deer** population' (17). Soon afterwards we find out that lynxes killed 20 per cent of **roes** (*sarny*). The English reader may conclude that wolves eliminated 10 per cent of all species of deer living in the Białowieża Forest. However, the original tells us only about **red deer**. Mistranslation is, in this case, a presentation of false statistics.

d) 'Na naszych oczach giną leśne **kuraki**' (89) – 'Before our eyes vanish **forest cocks**' (89). As explained elsewhere, *kuraki leśne* are **woodland grouse.** Obviously, the term refers to both cocks and hens. The translator was probably distracted by the gender of the word *kurak* in Polish, which, naturally, has nothing to do with the sex of entities in the real (as opposed to linguistic) world.

f) 'Bywa że niektóre osobniki spędzają zimę samotnie korzystając w tym czasie z karmy wykładanej w niewielkich **paśnikach** dla **sarn i jeleni**' (111) – 'Sometimes certain animals spend the winter in solitary [sic], taking advantage of the fodder provided in **the small pastures** for **deer or roe**' (111). There are two problems of interest to us in this fragment. Firstly, ***paśniki*** in this context are not small pastures but **feeding racks**. Although it is true that the word refers to both, when we use the preposition 'w' we mean a small wooden structure, where fodder is put in winter for wild animals. In the sense of 'pasture' the preposition required is 'na'. Secondly, we must remember that roe deer is a deer, therefore there is no point in writing 'deer and roe'. Either we write **'red and roe deer'** or simply, deer, because no other species of deer currently lives in the Białowieża Forest. This strange distinction could also be attributed to the insufficient knowledge of the translator.

g) Wojciech Sobociński is also the author of the book ***Torfowiska*** *z bliska*, which is translated as *The* ***moors*** *nearby* (143). Those expecting the book to be about moors will be surprised – it concerns peatlands.

XXII. Topograficzna Mapa Polski 1: 100000. 1997-1999:

The whole range of maps covering Poland in fragments translates (the legend) *Ur.* standing for ***uroczysko*** as '**desert**'. Although one can imagine that a particularly remote and beautiful patch of desert can be called *uroczysko*, there are no deserts in Poland. Admittingly, there is Pustynia

Błędowska (the Błędowska Desert), an expanse of sand covered by scanty vegetation, but it is not a desert in the sense of what one can see in the Sahara, in Nevada or the heart of Asia. The places called *uroczysko* are usually fragments of forests somehow distinguished in the eyes of the locals (e.g. by being inaccessible or by possessing exceptional features) or mineral swells among swamps, which naturally stand out. The editors of the map series are guilty of serious misinterpretation.

4.2.2.1.1. *Pars pro toto* or overspecification

Within the type of misread syntagms an interesting subtype emerges – the erroneous use of a lower-rank category instead of higher-rank one.

I. Durrell (1974: 176) – Durrell (2002: 117):

Dormouse – *koszatka*
As elaborated in Chapter 3.3., under the heading 'Dormice', *koszatka* is but one of many dormouse species.

II. Durrell (976: 211) – Durrell (1999: 211):

Fritillaries – *perłowce malinowce*
In fact, 'fritillaries' refers to a group of butterflies (*dostojki*), of which *perłowiec (dostojka) malinowiec* (silver-washed fritillary) is but one representative.

III. Fowles (1975) – Fowles (1994):

a) 'Downstream the dim rush of the weir, and a hidden **warbler**; a rich, erratic, un-English song.' (235) – 'W dole rzeki stłumiony szum jazu, niewidoczna **piegża**; zestrój dźwięków chaotyczny, nieangielski.' (253)
The choice of *piegża* as the equivalent of 'a warbler' is indeed unfortunate. While it is true that *piegża* belongs to the warbler family, it is one of very few members that has not got 'warbler' in its English name – it is called 'lesser whitethroat'. Moreover, its song – a short dry rattle on the same note cannot be characterised as 'rich' and 'erratic'. The selection of warbler species whose song can be described in this way is long and one could even single out an 'un-English' species, i.e. occurring in France, where the story is set, but not in England. Whichever warbler sings, the author does

not choose to be specific. He just says ‘a warbler’ – **‘jakaś pokrzewka’**. That might imply ‘lesser whitethroat’, though hardly so, as well as the whole plethora of warblers endowed with rich voices, ranging from the commonest garden warbler (*gajówka*) or blackcap (*kapturka*) to the rarer Dartford warbler (*pokrzewka kasztanowata*). In addition to all that, the translator transposes the description of the bird’s song to the impression made by the simultaneous running of the song and ‘the dim rush of the weir’. Thus in the Polish version this is not the song that is rich, erratic and un-English but the song thrust against the backround of the noise produced by rushing waters. In a sense such an interpretation justifies ‘piegża’ with its dull and simple song. But it is not what the author wrote and what feeds itself into the imagination of the readers of the original. Therefore this error could belong to the category of misinterpreting scenes/scripts, too.

b) the **owl** (264) – ***puchacz*** (274) – replacing the general with the specific.

IV. Fowles (1977) – Fowles (1982):

‘It was the world before the machine, almost before man, and what small events happened – **the passage of shrike**, the discovery of a new path, …’ (51) – ‘... – **przelot srokosza**, …’ (60)

Srokosz is ‘great grey shrike’ and is just one of many shrikes (e.g. red-backed or lesser grey shrikes, etc.). In Polish this group is named ***dzierzby***. Some of them, like ***dzierzba czarnoczelna*** (lesser grey shrike), have *dzierzba* in their name, while some others, like ***srokosz***, do not. Nevertheless, in generalisations both ***dzierzba czarnoczelna*** and ***srokosz*** are covered by the name ***dzierzba***. In English the situation is much clearer because all the European shrikes have ‘shrike’ in their names. In conclusion, the translation of the umbrella term ‘shrike’ as the specific *srokosz* is unauthorised, especially in the Mediterranean context, so rich in shrike species.

V. Grootjans and Wołejko (eds) (2007: 29):

Torfowiska alkaliczne **– chalk mires**

Alkaline does not mean ‘chalk’. Although these fall within the ‘alkaline’ category, they are not alone. One of the co-hyponyms unjustifiably functions here as a hypernym.

VI. Hryniewiecki and Okołów (2004: 26):

Brusznica is correctly translated as cowberry but on page 29: 'Liczne są krzewinki – **borówki**, malina kamionka ...', a hypernym is rendered with the help of a lower-rank category: 'Shrublets are abundant such as **cowberry**, stone bramble ...'

VII. Kossak (2001: 127):

'There was literally no river or stream, lake or **peatbog** without their characteristic lodges …' In the Polish text, after 'lake' we read '**torfowiska czy bagna**', which expresses more than the English text implies. *Bagno* is not necessarily on a peat substratum, *torfowisko* definitely so. 'Peatbog', as the name indicates, has peat and of a particular type – very acidic, very poor in nutrients. Such peatlands differ from other types in the way they look, so the Polish reader receives a much wider inventory of images than the English one.

VIII. Miłosz (1989) – Miłosz (1981):

a) ***pardwy*** (216) – **ptarmigan** (238). The latter is just one of the three species called *pardwa*, which in fact happens not to occur in the place described by Miłosz (see: 4.2.3.3. Insufficient knowledge of the subject-matter).
b) unspecific ***mysz*** (177) (mouse) is rendered as 'field mouse' (195) (*mysz polna*).

IX. Sawicki (2003a):

Brodźce (89) – **redshanks** (90). Only one species of *brodźce* is called 'redshank' in English, namely *krwawodziób.* Most appropriately, they are *Tringa* sandpipers.

4.2.2.2. Misinterpreting scenes/scripts

In fact, such a misinterpretation usually flows from the misreading of syntagms and verb frames, so frontiers between these subcategories are particularly fluid.

I. Durrell (1975a) – Durrell (1991):

a) ' … behind the house, where the forest began, a tinker-bird was giving its monotonous cry, toink… toink… toink, **like someone beating forever on a tiny anvil** ...' (13) – '… krzyczał monotonnie rudzik, zawodząc toink… toink…. toink, **niczym ktoś, kto niesłusznie dostaje lanie** ...' (14)

   Apart from the wrong translation of the bird's name (it should be *wąsaczek* – the author does not specify which species of the whole family of *wąsaczki*), the comparison to someone who is being spanked for no reason digresses greatly from the original. This is groundless, especially as the onomatopoeia remains transcribed in the same way. Everyone can see such sounds cannot imitate moaning. The translator evokes a totally different scene – beating somebody rather than beating on a tiny anvil.

b) '… a black-and-white wattled plover that **flapped away upstream**, tittering imbecilically to itself, with **its feet grazing the water**, and long yellow wattles flapping absurdly on each side of its beak.' (28) – 'czarno-białe siewki, **człapiące w górę strumienia** i chichoczące idiotycznie do siebie, **nogami grzebały w wodzie**, a ich długie, żółte korale trzepotały absurdalnie po obu stronach dzioba.' (37)

   The scene painted by the translator is rather strange: some sort of unspecified plovers walk with heavy steps up a stream scratching the water in the way chickens scratch the ground. Not only is the description unfaithful to the original but also to what potentially can physically happen. For analysis see page 221 below.

II. Fowles (1977: 50) – Fowles (1982: 59):

> 'Its [Phraxos'] beauty was rare even in the Aegean, because its hills were covered with pine trees, Mediterranean pines as **light** as **greenfinch** feathers.'

> '… śródmorskie pnie o kiściach igieł tak **lekkich** jak pióra **dzwońca**.'

Fowles likens the colour of the pines' needles to the colour of greenfinch feathers. Contrary to the translator's implication the bird's feathers are not exceptionally light in respect of their weight but are definitely light green and it was this feature that the author explored. How blind she was when choosing to rely on the script 'comparing pines and greenfinch feathers in respect of weight' is proved by her attempt to rationalise the comparison.

To that end she writes not just about pines themselves but the clusters of their needles.

III. Fowles (1978) – Fowles (2000):

a) 'The rabbits running, the **greensward they had won from the bracken**.' (379) – 'Szelest królików **szukających w paprociach trawy**' (455). The Polish version implies that under the bracken grows grass. The bracken is a very invasive species of fern shading the ground below its fronds so effectively that no light-loving plant can survive. The English text says something radically different – the greensward has been formed thanks to rabbits, which by grazing there regularly have prevented the bracken from spreading. The translation should then be as follows: 'Szelest królików, zielona **murawa, dzięki nim nie zarośnięta orlicą**'. The script is not 'rabbits looking for grass' but 'rabbits feeding on the grass they have helped to grow'.
b) 'He went to the window and stared down at the rivulets of water on the drive up to the house, at the **snowdrops** that clustered around the two old rick-stones that bordered the path of the front door' (450) – 'Stanął przy parapecie i zaczął się wpatrywać w strużki wody, spływające podjazdem, w **resztki śniegu**, które zebrały się wokół dwóch starych kamieni na skraju ścieżki do domu.' (536)
One can easily imagine patches of snow around stones early in the spring but the context tells us that for quite some time already it has been warm and rainy, without a single snowflake in sight. Moreover, snowdrops is the name of the popular early spring flower (*Galanthus nivalis* – ***śnieżyczka przebiśnieg***) and the normal way of describing snow is with the help of the word 'patches' not 'drops'. The translator misread the syntagms and imagined a completely different scene, following a typical (for Poland rather than Britain) winter-and-early-spring-thaw script.
c) 'Then Dan was given an ornithological treat. He looked through his field glasses at a **robin-like** bird hopping in the shade by the river's edge. It was a very handsome little creature, **a bluethroat**, **the first of the species** he had ever seen.' (563) – 'Dan otrzymał za to ornitologiczną nagrodę. Przyglądał się przez lornetkę **podobnemu do drozda** ptakowi, który uwijał się, podskakując w cieniu na brzegu rzeki. Był bardzo piękny, **z niebieską szyją**, Dan **nigdy jeszcze takiego nie widział**.' (667)

The translator ignored two clues included in the context which should have helped him translate 'a bluethroat' as ***podróżniczek.*** Firstly, the words 'the first of the species' suggest that 'a bluethroat' is the species' name rather than a brief description (in the Polish version 'z niebieską szyją' means 'with a blue neck'). Secondly, the bird was similar to the robin. Here, the clue was not disregarded but misinterpreted – the bluethroat is not thrush-like (*podobny do drozda*) but robin-like (***podobny do rudzika***). In effect, the Polish reader is not invited to appreciate Dan's luck. For a hobbyist seeing a bird, not the particular species, is not an ornithological but merely an aesthetic treat, much easier to experience. The translator relied on a script 'admiring a beautiful sight'/'experiencing delight at encountering something charming', whereas a more focused one should have been evoked – 'experiencing joy/pride at encountering a species of bird a birder has not seen before'. Alternatively, the error could be categorised as an unjustified employment of hypernymy (meta-translation errors).

d) 'The route lay over long and treeless uplands, a barren and color-drained moor. Solitary lapwings stood by mournful **pools**.' (621) – 'Jechali długo przez wyżynne i bezdrzewne pustkowie, jałowe i bezbarwne. Samotne czajki sterczały nad posępnymi **stawami**' (732).
The image in the Polish version, with the ponds in it is quite different. The word ***stawy*** implies man-made structures looked after by people. The English text paints a wild, rough landscape with no obvious human presence and associated orderliness. Therefore, pools must be either small lakes (*jeziorka*) scattered around the moor or puddles of rainwater (*kałuże*). Considering the weather described earlier and the place (Syria), the latter looks like a safer choice.

IV. Kłosowscy (1999):

a) Te torfowiska, zwane wysokimi, pokrywa bujny, wybrzuszony kobierzec mchów, a latem **bielą się** one puszystymi owocostanami wełnianek.' – 'These so-called highmoor peat marshes are covered with a rich, bulging carpet of moss, and in the summer they **bloom** with fluffy cotton-grass fruits' (10). Too often are the fruiting heads of cotton-grass taken for its flowers. As we can see, this deeply-rooted and widely held belief is reflected here in the choice of a rather oxymoronic collocation 'bloom with fruits'. In the original the fruits '**gleam with whiteness**', not the flowers. According to the most typical script, it is the flowers that are white, hence the error.

b) 'A caption under the photograph showing a **scree** (in Polish it is 'Na **gołoborzu**') reads: 'On **bare slopes**' (275). While the message in English is not incorrect, it is more descriptive and less technical. If it was not for the picture, one could imagine a smooth slope without piles of loose rocks, which form a scree. The translator must have blindly stuck to the scene implied in the Polish term by *goły – bare*, meaning 'without' and *bór* – in folk understanding: forest, forgetting about the fact that *gołoborze* is a specialist term.

V. a) In the passage cited beneath the translator relies on the same scene for *uroczysko* and *ostęp,* though they do not always overlap, differing diametrically from context to context. By using the same word to translate the two words and, additionally administrative units *oddział* and *straż*, she blurs the distinctions between formal/informal denotations and between divisions for hunting purposes and forest management. In Kossak (2001: 130) ***ostępom*** is rendered as 'forest ranges': 'the Hunting Office has put forward an objection that the mines do harm to **forest ranges**, beaver colonies, ...'. The use of the word 'range' is quite confusing here, because most of the time it is given the meaning of ***uroczysko***, the opposite of what *ostęp* stands for. The matter is further obscured a few lines below when the territorial unit for gamekeeping (***straż***) is also called '**range**'. If we recollect that on page 98 '[t]he bear site was arranged in the **range** No. 317...' – '... w **oddziele** 317 ...' and that in Pudlis (2003) 'range' denotes *obręb*, we come up with five different meanings, whose semantic fields either partially overlap (*oddział, straż*) or exlude each other (*ostęp* vs. *uroczysko; obręb* vs. *oddział).* It is worth noting that *uroczysko* usually comprises *ostęp* 'secluded, wild place', for example on page 141, we find out that in order to create a vast game enclosure: 'A few hundred hectares of pristine forest in **the Wielka Kletnia range ('na uroczysku zwanym Wielka Kletnia')** near the Jelanka (now Orłówka) River source were fenced off with a huge oak palisade'. However, *uroczysko*, though situated deep in wilderness and containing inaccessible sites, is not reserved for wildlife but is managed for human purposes as a whole.

b) 'you can hear the loud banging and cracking of antlers'. The Polish reader receives a slightly different description of what can be heard: 'gniewne uderzanie wieńcami o drzewa lub trzask posuszu pod nogami' (201). More details are brought to the fore: antlers are banging against a tree, brushwood is cracking under the hoofs. The translator brought into view a different script: 'noises accompanying the fight of two stags' instead of 'noises produced by stags preparing for combat'.
c) It sounds rather odd when amidst a passage about natural regeneration, the following sentence appears: 'If a lot of seeds **are planted** just after a fire, young pines soon replace burnt spruces' (232). The correct rendering of 'Jeśli więc rok obfitego obsiewu zbiegnie się z terminem niedawnego pożaru, …' should be 'If then **the year of seed abundance** comes just after a fire incidence …'. Seeds do not need to be planted, they can germinate with the help of natural conditions such as opening up a canopy of dense forest by burning trees and shrubs. 'Obfity obsiew' cannot be rendered in the passive voice.
d) 'The cock [of a capercaillie] furled and unfurled its beautiful tail like a lyre' (474) – Lyre-shaped tails are characteristic of black grouse, not capercaillies. This is a harmful overinterpretation of 'wspaniały wachlarz ogona', which is 'a magnificent fan of the tail'. The translator mixed up scenes – the performance of the lekking black grouse and capercaillie.
e) 'Bison prefer to graze in **fresh humid mixed forests**, …' (509) suggests that all the three conditions for the forests have to be fulfilled simultaneously if they are to attract bison. They must be fresh, humid and mixed. The Polish text does not imply anything of that kind. In it bison prefer to graze in **fresh broadleaved forests, wet broadleaved forests and mixed forests** (***w lasach świeżych, wilgotnych i mieszanych***). The forest does not need to be mixed and it does not have to be wet in order to host bison. The scene envisaged by the reader of the target language text diverges from the source significantly.

VI. Lewandowska and Lewandowski (eds) (1994: 40):

The caption next to a photograph of a song thrush fledgling perching on a branch is '**Podlot** drozda śpiewaka' – 'A song thrush's **flight**'. The contrast between what can be seen and the description of it immediately brings a wry smile to one's face. ***Podlot*** is a **fledgling** – a young bird fully covered with plumes but not yet able to fly well. The translator has apparently interpreted it as ***podlatywanie*** – 'trying to fly' or 'flitting from branch to branch'.

VII. Miłosz (1989) – Miłosz (1981):

a) '… tylko ramiona krzyża zakryte miał liśćmi **barwinka**.' (28) – 'Jesus' body […] that normally hung on the crucifix, only now the arms were covered with **myrtle** leaves.' (28).
   Barwinek is 'periwinkle'. Anioł-Kwiatkowska (2003) does not even mention any other possibility. It is a common practice to plant periwinkles around roadside crosses and shrines in Poland and Lithuania, where the book is set. Here, the cross is adorned with its leaves. The myrtle occurs naturally neither in Poland nor in Lithuania and is not grown in gardens due to too harsh climatic conditions. Admittedly, the myrtle is known (e.g. thanks to the Bible) as a decorative plant with religious implications in Mediterranean countries, which could have automatically influenced the decision of the translator. One can imagine that also the readers might quite automatically accept that the leaves used for decoration were those of the myrtle until some of them realise that bringing the plant to the far-flung northern province and its survival there is not plausible.
b) 'Polowanie w rojście różniło się ostrożnością – a nuż która siedzi na krzaku **bahunu** czy między jagodami **pijanic** i tym, że miękki mech nie pozwalał jej porządnie ogłuszyć, ...' (121) – 'Hunting in the swamp, where every **bilberry** or **rosemary bush** might conceal an adder, required even greater stealth – and skill, because the soft mossy matting cushioned every blow of the stick.' (133)
   Unfortunately, this time translation is so inadequate that the clues given by Miłosz are lost.
   Bilberries grow in much drier places than bog bilberries, which are the equivalent of *pijanice*. The rosemary bush, whose leaves as a matter of fact do resemble those of the marsh tea (*bahun = bagno zwyczajne*), is a Mediterranean plant, which needs a lot of sunshine and not much water. It would not survive in a damp bog, additionally shaded by pines. Thus, the swamp of *The Issa Valley* is a peculiarly dry spot, unlike *rojst* of *Dolina Issy*. It cannot even vaguely refer to the raised bog, which it must be according to the vegetation described by Miłosz. 'Swamp' in itself does not fit the description – 'bog' should replace it. Strangely enough, ***bahun*** is

translated correctly as '**marsh tea**' and ***pijanice*** (1989: 235) are '**bog bilberries**' (1981: 214).

Admittingly, Americans use 'wild rosemary' – but not 'rosemary' on its own – as a synonym for the marsh tea (in Blamey, Fitter, and Fitter 2003 – the Labrador tea).

c) 'majaczyły **karłowate sosenki rojstu**' (157) – 'loomed **a wall of pine scrub'** (173)

'Pine scrub' suggests the pines grow densely and, potentially, on dry ground. Here, the pines are scarce, dwarfed and sticking out of a moss-cushioned bog. The translator concentrated on the scene featuring 'karłowate sosenki' so much that 'rojstu' escaped his attention – the latter, meaning 'bog', can barely support 'scrub'.

d) '**tokuje** jedynie o świcie' (154) – 'It [capercaillie] hooted only at daybreak' (170)

Tokować means both utter courtship voices and display in front of females in a variety of ways. The English equivalent embracing both elements is 'to lek'. 'Hoot' seems to refer only to the vocal part and in the case of capercaillie it is extremely misleading since it suggests that the bird hoots like owls or produces a loud raucous noise, while in fact its song consists of fairly quiet clicks, wheezes and a pop.

'**toki** są akurat w pełni' (157) – 'the **hooting** was nearing its peak' (173)

'**Dźwięk** odzywał się teraz już wyraźniej. Jakby wydzierane z wysiłkiem brzmiało tek-ap, tek-ap, coraz prędsze.' (158) – 'The **hooting** carried more distinctly now – a strenuous, quickening *tek-ap, tek-ap*.' (174)

'bo ten **tokował** teraz, zdawało się, bardzo blisko' (159) – 'so close was the **hooting** now' (175)

This description, together with the details soon to follow, has not persuaded the translator to give up the employment of 'hooting'. What is more, he seems to insist on inserting the word even where it is not necessary:

'Tek-ap, i coraz prędzej, coraz prędzej, aż nowy odgłos się dołączył – jakby ktoś ostrzył nóż ...' (158) – '*Tek-ap* ... . Faster and faster, until **the hooting** was joined by another sound, rather like knife-grinding' (174)

Here it is really undesirable because one may conclude that the bird starts its song with hooting and then some other sounds join in. How wrong such a conclusion would be is shown by the words

written about the beginning of the performance on page 173: 'At times it sounded vaguely like a **hammering**, an **uncorking**'
Despite everything, the translator manages to render ***toki*** (1989: 177) in a far better way:
'cietrzewie upodobały sobie to miejsce na **toki**' – '... the blackcocks had chosen as the place for their **mating ritual**' (1981: 161)

e) 'wijące się korzenie zwalonych drzew, plątawisko nagiej łozy, łomu i wykrotów' (157) – 'the tangled underside of an uprooted tree, clumps of **rushes**, **brake**, and **sedge**' (173)
The picture painted in the Polish text is this: 'creeping roots of wind-blown trees, tangled osiers, dead branches and fallen trunks'. The published English version introduces a totally different scenery – there are rushes (*sitowie*) characteristic of edges of lakes and other water bodies and sedges (*turzyce*), abundant enough to deserve mentioning, which suggests the landscape is open (in forests sedges rather than blanketing large tracts of land are not very conspicuous). The most mysterious of all is the **brake.** It can either refer to bracken, a huge invasive fern (*orlica*) or to any area of brushwood, thick undergrowth or thicket of one kind of plant. The landscape in the English text is a rich mosaic of various habitats, while in the Polish original, the focus is on just a fragment of bog woodland. How the translator managed to imagine a scene so distant to the original one remains a mystery.

VIII. Sawicki (2003a):

The Polish sentence 'Rybitwa czubata, jak sama jej nazwa wskazuje, ma z tyłu beretu sterczący czubek z dłuższych piórek' (96) is translated as 'A sandwich tern – as its name indicates – has a sticking tuft of longer feathers at the back of its beret' (97). That the English name, contrary to the Polish one, does not suggest anything of that sort escaped the translator's attention. The image imposed by the native tongue name of the tern has paralysed common sense, which should have made the translator realise that 'sandwich' could not possibly evoke the image of a taft of feathers on the bird's head.

IX. Sidło, Błaszkowska and Chylarecki (eds) (2004):

Olsy (231) – **riverside carrs** (233)
The context is that these forests grow in lowest-lying wetland areas in river valleys and around lakes. Certainly, this is a typical site for riverside carr, but in the original we are told about alder swamp forests occupying those areas. The translator clung to the typical scene, overlooking subtle but pivotal distinctions in its composition.

4.2.2.3. Misconstruing the text's modality

The recognition of the text's modality consists in discerning the author's intentions, tones, attitudes, views or perspective. When the translator ignores linguistic markers which allow the reader to notice all these, errors creep in.

I. Fowles (1978: 352):

Here is a comment on the scene in the States (in Tsankawi) where ravens were mobbing a red-tailed hawk, typical behaviour of many smaller species against raptors:

> This minor coincidence of seeing two species and a common behaviorism of the English countryside reproduced in a very foreign and remote one, somehow seemed to prove it; all my lasting relationships were with this world of quasi-arcane knowledge and experience ... not merely of course in natural-history sense, but because I was fundamentally an observer and storer of correspondences …

Later in the book this observation is well exemplified:

> A hidden magpie chattered from the far side of the combe, and there was an angry cawing from a pair of crows. ... It was a buzzard, circling high over the beechwood, the soft light from the opened west catching, as in some gentle, delicate searchlight, the brown-and-white underside of the bird's flexed wings. It mewed, majestic, golden, apotheosized, against a dark cloud above. Dan stood and watched until it was **chased off**, remembering Tsankawi. (456)

The Polish translator has disregarded the apparent correspondence between two events in two different places. He has overlooked the importance of the workings of nature to Dan and the author himself and that is why the last sentence reads: 'Dan wpatrywał się w ptaka, dopóki ten

nie odleciał, i wspomniał Tsankawi' (Fowles 2000: 543). In this fragment Dan watched the buzzard until it flew away rather than was chased off by the crows. The close and well-specified link between the scene in Tsankawi and the scene in England is not marked. The readers are left less guidance and more freedom in interpreting the text. In other words, they receive a text diverging from the original in which the author clues us in as to his interpretive intentions sometimes through hints and sometimes directly.

II. Kłosowscy (2005: 172):

Although we talk of text's modality mainly in reference to belles-lettres, it seems to me, the following example from outside it is worth considering. After 'The roller, only 20 years ago a common inhabitant of the eastern Poland, for unknown reasons has become increasingly rare' (the grammar as in the original). The following comment appears in the Polish version: '**Jest to tym bardziej niezrozumiałe**, że nie unika sąsiedztwa człowieka, gnieżdżąc się w dziuplach drzew stojących na skrajach wsi.' This is rendered as '**Interestingly**, it does not avoid human vicinity, nesting in tree holes on the outskirts of villages.' In Polish, the fact that the roller mixes well with people and yet its numbers keep dwindling is presented as something that adds to the incomprehensibility of the disastrous trend. In English, this situation is just interesting, worth noting. The message is not entirely wrong but is much less alarmist.

III. Kossak (2001: 527):

Przestój (in the plural: *przestoje*) is overstaying tree = a tree which was not felled in due time, according to the felling rotation planned for a given stand or species. On page 538 '"przeszłorębne przestoje"' becomes 'superannuated trees'. Whichever the translation, the most interesting aspect is that the same kind of tree is called by Polish nature conservationists ***pomnik przyrody*** **(a monument of nature)**. In English they are often referred to as **veteran** or **ancient trees.** What a different attitude these category labels encode! On page 357, the translator has chosen not to convey the author's distance, or veiled sarcasm stemming from putting in converted commas ***przeszłorębne świerki*** and written: 'As the quantity of deadwood was growing with years, the permit to cut trees of various species with dry tops, dying pines and **mature spruces** of 49 cm at breast height ... was granted ...'. 'Mature spruces' without quotation marks gives the sentence the flavour of a dispassionate account.

V. Miłosz (1989): – Miłosz (1981):

a) 'dywanik z **łosia**' (94) – 'an **elkskin**' (104) This would be fine if not for the fact that the translator writes in American English and, for example, uses the American name 'kinglet' for European relatives of the bird, called 'goldcrest'. In American English *łoś* is 'moose'. This lack of consistency is confusing, which does not endorse 'an americanisation' of European species. The use of names of American relatives of European birds shifts the author's perspective and shades his intentions of returning to childhood days in a place so alien and peculiar to westerners.
b) To preserve the text's modality, the technique of domestication, especially American domestication, should be approached with utmost caution:
'Wpełzał w cień pod krzakami, gdzie nie ma trawy tylko splątane pokłady suszu, a między nimi **mysikróliki** przemykają z trwożliwym czik-czik-czik' (59) – 'he crawled into the shady underbush, among matted layers of dead leaves, where **kinglets** flitted with a tremulous cheep-cheep-cheep' (64).
Kinglets are American species – *Regulus satrapa (mysikrólik złotogłowy), Regulus calendula (ogniczek).* They are very similar to the European *Regulus regulus,* called in English '**goldcrest**' and in Polish, *mysikrólik.* However, they do not occur in Europe and it is really strange to find their names in the description of native European nature.

4.2.3. Realisation errors

Such errors appear when the translator, having come up with a suitable interpretation of a text, fails to find ways to ensure that the recipients arrive at it as well.

4.2.3.1. Target language errors

I. Chylarecki and Sawicki (2003):

a) The translator comes up with his own name for the parti-coloured bat (*mroczek posrebrzany)* – 'batline' (41). Although the authors mention that the species is in the *Polish red data book of animals,* he has apparently failed to consult the very source, where the correct English name is given (cf. Głowaciński (ed.) 2001: 55) .

b) 'Po tym okresie z jaj wylęgają się pisklęta, które w odróżnieniu od piskląt gołębi, sikorek czy wróbli, wykluwających się jako ślepe, nagie i niedołężne, są **zagniazdownikami**, tak jak dobrze nam znane kurczęta.' (44) – 'Then hatch the **squealers** which, unlike the blind, bare and helpless nestlings of pigeons, tits and sparrows, present a development strategy of newly hatched chickens' (45). ***Zagniazdownik*** is in English a **precocial bird,** helpless nestlings of pigeons, tits, etc. being called altricial birds (*gniazdowniki*). 'Squealer' signifies: 'one that squeals: as 1. A European swift; 2. Harlequin duck; 3. An American golden plover; 4. A young squab pigeon; 5. a young grouse, partridge or quail; 6. informer, betrayer '(Webster's). 'Precocial bird' is too remote in meaning to be among the possibilities. The use of the word 'squealer', which can just as well refer to altricial birds, makes the message a contradiction-in-terms. The interpretation of syntagms may not have been incorrect since the authors explained everything plainly but the choice of word was unfortunate due to the insufficient knowledge of target language terminology.

II. Durrell (1975b: 214) – Durrell (1999: 227):

> I have met a number of people who professed to be interested in birds, and who had turned out (after careful questioning) to be charlatans who did not know what a hoopoe looked like, or could not tell the difference between a **black redstart** and an ordinary one.

'Black redstart' becomes ***pliszka kaukaska*** in Polish instead of ***kopciuszek***. 'An ordinary one' refers to 'redstart' – ***pleszka***. Since in Polish the two species do not share any common element in their name, the translator has to write '... pomiędzy **kopciuszkiem** a **pleszką'** (' ... between a black redstart and a redstart'). Deciding on ***pliszka kaukaska***, the neglectful translator brings shame on the author, who is expressing his dissatisfaction with those that feign interest in birds, and himself appears incompetent. The error is particularly malicious because in Durrell (1999: 112) (in English – Durrell 1975b: 106), in a context where it is definitely less conspicuous, 'redstarts' are ***pleszki***. The translator may have been misled by the adjective 'ordinary': if the redstart is so ordinary, the black redstart must be extraordinary in one way or another. Dissatisfied with the plain-sounding *kopciuszek* (when the first letter is capitalised it means Cinderella), she went for a more exotic-looking *pliszka kaukaska* – the Indian pied wagtail, which is not even related to both species in question but deceptively suggests taxonomic closeness (*pleszka* – *pliszka*). Whatever the motivation, the target language terms are muddled.

III. Durrell (1976: 143) – Durrell (1999: 30):

Red cardinal – ***rusałka admirał***. The red (northern) cardinal is a bird, ***kardynał szkarłatny***, *rusałka admirał* is a butterfly called 'red admiral' in English, which the translator realises on page 212. Most probably, she absent-mindedly put down *admirał* instead of *kardynał,* as both words sound similar and invoke associations with high-rank officials.

IV. Fowles (1977: 299) – Fowles (1982: 358):

Exactly as in Ia) above, in the following fragment, the translator ignores existing names, introducing neologisms or obscure archaisms. **'Smew'** and **'goldeneye'**, which are mentioned by an ornithologist – a specialist after all, become ***szlacharki*** and ***podgorzałki krzykliwe*** instead of ***bielaczki*** and ***gągoły***. Whereas ***szlacharek*** denoting the smew can be traced back to a source (*Rok myśliwego* by Włodzimierz Korsak, published as long ago as in 1922), ***podgorzałka krzykliwa*** is nowhere to be found. Certainly, guide books on birds published over the last thirty years mention neither. The closest one can get is ***kaczka krzykliwa***, the creation of an early Polish naturalist, Stanisław Jundziłł (1761-1847), but according to Strutyński (1972), *gągoł* has always dominated as a local name and in the works of Polish zoologists (77).

V. Fowles (1978) – Fowles (2000):

a) Fowles tells us that 'a stook [is] four pairs of sheathes and a single "to close the door" at each end' and that to stook is 'to stand sheaves propped against each other, setting them, shocking down the butts into the stubble and simultaneously, clashing the eared heads together' (2). It is absolutely clear that stooking is 'ustawianie **kopek/styg** zboża'. Yet in the Polish version the process is called 'ustawianie **stogów**' (8). A prototypical *stóg* is a pear-shaped heap of hay formed around a tall wooden pole (= a rick). The word can also refer to a large heap of straw – stems without grain. Both constructions are designed to last several months, stooks being short-lived items collected after a few days. Without much risk we can say that the target language verbalised categories of *stygi* and *stogi* merged into one when it came to the final draft.

b) In the description of the cornfields, Fowles tells us about 'seeding **thistles**'. In Polish, we have 'sypią się **głogi**'. *Głóg* – hawthorn – is a small tree or a shrub and its fruit does not even distantly resemble thistledown that features in the description. The same mistranslation can be observed later as well: 'he went up on the tractor behind Louise with the hay-mower to cut the top ley clear of its nettles and **thistles.** He followed after her with a pitchfork, heaping the cut stems for later burning.' (Fowles 1978: 374) – '... usiadł za Louise na traktorze, do którego przyczepili kosiarkę i pojechali oczyścić górny ugór z pokrzyw i **głogu**. Szedł za maszyną z widłami, układając **ścięte łodygi** w stosy przeznaczone do późniejszego spalenia.' (Fowles 2000: 448). If the hawthorn rather than the thistle grew there, the hay-mower would have to be preceded by somebody with an axe. Despite the fact that the translator chose to write ***łodygi*** (used to refer to non-woody stems) rather than ***pnie*** (referring to tree trunks) or ***badyle*** (indicative of bush stems), he did not notice his mistake, so well-imprinted must have been the conviction that *głogi* was the same thing as *osty* (thistles). Interestingly, later in the book (e.g. Fowles 1978: 378 – Fowles 2000: 453), the word *głóg* is used to denote what it really does. Unfortunately, this realisation did not stimulate any reparative action in relation to the earlier contexts.

c) 'A large flat-topped limestone rock stood there, the "Pulpit", isolated in the sea of **bracken**. ... Daniel and she stood there, chest-deep in the **fronds**. Her elder sisters had brought her there when she was little, they used to clear a "room" in the **ferns**.
Now they did the same, squashing the stems sideways with their feet, treading the stiff ends flat, till there was a close green-walled room again, six by four.' (387) – 'Sterczała tutaj płaska wapienna skała, Skalny Pulpit, otoczona morzem **paproci**. .... Stanęli po pas w **zaroślach**. Kiedy była mała, przyprowadziła ją tutaj starsza siostra; razem wydeptywały pokój w **chaszczach**.
To samo zrobili teraz; nogami rozgarniali łodygi na boki, uparcie deptali sterczące końce, aż wreszcie mieli izdebkę sześć na cztery, otoczoną ścianami zieleni.' (464)
The English reader imagines the whole scene among the ferns. The two children are obviously strong enough to make a "room" in the bracken by squashing its stems. The images created in the mind of the Polish receptor are quite different. The youngsters stand in the

thicket of bushes (*zarośla*) and decide to make a "room" there by treading the thick and tangled bushes (*chaszcze*). In the next paragraph they do so surprisingly easily and the hard shoots suddenly become soft stems such as those of ferns. Lack of consistency makes the Polish text puzzling if not ridiculous. The translator does not realise that the category of *zarośla* and *chaszcze* in the target language suggests that most, if not all, stems are woody and tangled.

VI. Iwaszkiewicz (1987: 400/401) – Iwaszkiewicz (2002: 12):

Białodrzewy – white poplars (correct) but:
'krynica z **żurawiem**, z czterema **srebrnymi białodrzewami'** (425) – 'a spring with a **water pump** and four **silver poplars**' (43) is obviously anything but correct.
Firstly, *żuraw* is not a water pump (it is **a well sweep**), which rather belongs to the category of mistaken syntagms. Secondly, the poplars are **white poplars**, silvery in appearance. One word is missing, the other needs to be slightly changed in the target language to express the same as in Polish.

VII. Kłosowscy (2005):

a) **Trzciniak** to jeden z najpospolitszych mieszkańców strefy szuwarowej naszych wód, dający o sobie znać donośnym, terkocącym śpiewem. Nazwa ptaka nie zawiera przesady, trzyma się on bowiem wyłącznie gęstych i rosłych trzcin, do których łodyg przypina swoje koszyczkowate gniazdo.

   The pipit is one of the most common inhabitants of the reed zone of our waters, known for its loud, rattling song. The Polish name of the bird (*trzciniak*, the small reed) is true to its nature, since it stays exclusively in dense and high reeds, on which it hangs its basket-like nest.' (122).

Trzciniak in English is the **great reed warbler. The pipit** is **świergotek** and there are four different species in Poland: meadow (*łąkowy*), tawny (*polny*), tree (*drzewny*), and water (*siwerniak*) pipit. Unlike the first three *siwerniak* is not preceded by the word *świergotek*. None of them lives in the reeds, and all of them sing in a gentle, pleasing way. As can be seen, taking the great reed warbler for a pipit (which species?) is almost incredible.

b) ***Trznadel*** (shown in a photograph) – the caption reads 'the bunting' (167). Quite ironically, the yellowhammer (*trznadel*) is the only bunting without 'bunting' in the name. Besides, the buntings are a family within the suborder of *Oscines* (singing birds), but which of the buntings is **the** bunting cannot be determined.

'The' cannot appear either before 'pipit' or 'bunting' because these do not stand for a particular species but a group of them and as such they can take the definite article only when they signify the whole group in the plural. Interpreted in this way, the errors cited are realisation errors but the translator may have simply had insufficient knowledge concerning the names of species and, of course, in the case of *trzciniak* what happened primarily was the mistaking of syntagms.

VIII. Lewandowska and Lewandowski (1994: 42):

The caption to the picture on the opposite page showing a young roebuck is ***koziołek***, translated as a 'roebuck', but the caption to the photograph on the same page, depicting the head of a roebuck, which in Polish is ***kozioł*** (**old roebuck**), is a **'goat'**. It is true that *kozioł* means primarily 'male goat', but here, there is no room for doubt. The misreading of syntagms seems improbable. The translator could have subconsciously written the word 'goat', 'hypnotised' by the superficial appearance of the source language lexeme – *kozioł*, which derives from *koza*, the name of the species as such and of the female goat. After all, the male goat is 'buck'.

IX. Lewandowska and Lewandowski (1999: 36):

'Larwa **zawisaka gładysza'** (properly **'zmrocznika gładysza'**) is rendered as **'a hawk larva'** instead of an elephant hawk-moth larva. The hawk is a bird, ***jastrząb*** in Polish, and reading about its larva must make people realise there is an error but leaves them none the wiser as to the correct version. The decision to put this example in the category of target language errors is motivated by the improbability of mistaking a nestling for a larva; '-moth' must have vanished from the text by accident.

X. Miłosz (1989) – Miłosz (1981):

a) 'Latem noszono przeważnie łapcie, wyplatane z lipowego **łyka**' (6) – 'In summer **bast** moccasins from linden **bark** were worn' (5)

If the shoes were bast, they must have been made of bast, not the whole bark. The target language phrasing is muddled to the point of self-contradiction. It would look more logical as follows: In summer moccasins from linden bast were worn.

b) '… z **pagóreczka** w **dolinkę**, tu **borek**, tu rojścik ...' (114) – '... a **knoll** sloping down into a **glen**, a clump of **woods**, a patch of swamp …' (126)
The image evoked in the Polish speaker's mind is that of a low hill sloping down into a small, not very deep valley. The English 'knoll' is indeed a low hill with a rounded top and gentle slopes. In view of this fact it requires high levels of abstract thinking to picture how it manages to neighbour a deep, narrow valley, which is a glen by definition. The translator may not have realised the real features of 'glen', having in mind its vague gestalt and clumping it together with 'any valley'. As far as *borek* is concerned, the English version gives more choice to the reader in deciding upon what type of woodland appears, interspersed among the hills. In Polish, we must imagine small clusters of conifers.

XI. Puszcza Białowieska. Mapa przyrodnicza 1: 50000:

Świerczyna na **torfie** – spruce forest on **high moor**
'High moor' is probably *torfowisko wysokie*, which suggests a felicitous interpretation on the part of the translator. But the choice of the English term is rather unfortunate in Polish conditions (see for comment Chapter Three or 4.3.2.1. beneath). In order to avoid this awkwardness literal translation would be acceptable – spruce forest on **peat**. *Interpretation manual of EU habitats* EUR 27. 2007 encodes this habitat as 91D4: **mire spruce woods**.

XII. Sawicki (2003a):

a) **'mokradła** stałe i **okresowe'** (22) – 'permanent and **periodic swamps'** (23). 'Periodic' in reference to swamps is actually a contradiction in terms. Swamps are permanently underwater. We talk of periodic wetlands.
b) **'obszary porośnięte szuwarami'** (72) **– rush areas** (73). It sounds almost like rush hours. Such a collocation is impossible and must be replaced with 'areas overgrown with rushes'.
c) **'zimujące ptaki wodno-błotne'** (80) **– winter-abiding** (81) water and mud birds. These are wintering wetland birds. In order to

survive winter, they have to abide by the harsh rules of this season, but the phrase 'winter-abiding' is rather contrived.

d) 'Być może z gęstego **szuwaru** poderwą się kaczki' (88) – 'Maybe some ducks will start from **a rush**.' (89) Section 4.2.1.4. designates 'a rush' as a calque of *szuwar*, which both in the singular and plural denotes an uncountable mass of rush vegetation. It owes its presence in this section to being concurrently recognised as an instance of erroneous usage of the indefinite article.

XIII. Sawicki (2003b: 23):

> Bardzo cennym i unikatowym na tym terenie zespołem leśnym jest żyzna buczyna pomorska, której północno-wschodnia granica zasięgu przebiega przez Pojezierze Iławskie.

is rendered as:

> **The Iławskie Forests, in addition to various types of forests, have also quite often the beach areas. However these beaches are very rarely in their native and unaltered state.** The fertile Pomeranian beech forests, which has the northeast border line of its occurrence crossing the Iławskie Lakeland, is an is an especially unique and valuable.
> (all errors, including the repetition of 'is an' as in the original). (23)

The fragment which does not appear in the Polish version (in bold) is probably about beech forests, not 'beach areas' (why not beaches?) but the incorrect target language spelling and all the other errors (wrong word order, misuse of the article 'an' and the awkward phrasing of 'which has the northeast border line of its occurrence crossing the Iławskie Lakeland') evoke radically different scenes in the target language. The same can be classified as an erroneous addition (see below).

XIV. Sobociński (2004):

a) '… białowieskie **grądy, olsy**, łęgi, bory, a także nadrzeczne łąki i bagna naprawdę są ostatnią puszczą.' (6). – '... the **alder and oak woods**, riverside meadows and swamps really make the ultimate forest' (17). Two distinct types of woodland *grądy* – oak-hornbeam forests and *olsy* – alder swamp forests merge together, an interesting but unrealistic vision. The target language error consists in a failure to repeat the word 'woods' twice. 'Oak woods' and 'alder woods', though still not *grądy* and *olsy*, would not at least create a new reality, scenes imaginable in fantasy books but not in the Białowieża Forest. This error could then be classed as a misinterpretation of scenes/scripts, too. It can also fall into the other subcategory of realisation errors: insufficient knowledge.

b) 'Puszcza nie jest wyłącznie lasem'– 'The forest is not all but woods.' (37) Using the capital 'F' in 'forest' and replacing 'not all but' with 'not only' would eliminate awkwardness in this sentence. It would define the referent as the whole area geographically determined as Puszcza Białowieska. As such it comprises forest complexes, glades, meadows, marshes, human settlements and arable land.

c) '... aby żyjące tam rośliny mogły **wykiełkować** i **zakwitnąć** ...' – '... for the plants living there to **blossom** and **flower** ...' (51). This tautology in the English version is due to the wrong translation of *wykiełkować*, which is 'to sprout, germinate'.

4.2.3.2. Wrong evaluation of the recipient's knowledge

Hejwowski points towards the provision of insufficient information on culture-bound items as more widespread than offering too much of it. In the material gathered for the purposes of this study, this category of errors overlaps with those discussed in 4.2.4.6. 'Changing the text's intertextuality'. As far as nature terminology is concerned, one notorious case of exposing foreign readers to extra effort in comparison with the Polish recipients is providing Latin names of animal taxonomic families such as in Kłosowscy (2005):

a) *rodzina krukowatych* – the *Corvidae* (49) instead of corvids or still better – the crow family.
b) *ptaki siewkowe* – the *Charadriiformes* (76, 79) instead of the familiar waders or shorebirds.

4.2.3.3. Insufficient knowledge of the subject-matter

I. In the original (Durrell 1975a: 60) we read about dormice living in a giant hollow tree in an African rainforest and that they closely resemble our European dormice except that they are a pale ash grey with a slightly more bushy tail. The description is accompanied by the drawing on page 161. In the Polish version (86) '**dormice**' become ***susły*** instead of ***pilchy*** or ***popielice/popielicowate***, which is also reflected in the illustration on page 235. The translator overlooks the fact that 'sousliks' do not inhabit trees but burrow in the ground in steppes or open fields and their tails are never bushy. As two European species of dormice are grey, Durrell must have contrasted their pale ash grey African cousins with the ginger-coloured common dormouse (*orzesznica*) or garden dormouse (*żołędnica*). It does not change the fundamental matter that none of them resembles sousliks. The English original was published earlier, so the illustration which dispelled any doubts, was already available for the translator.

II. Fowles (1978) – Fowles (2000):

a) 'A buzzard **mewing** somewhere out of sight' (388) – '**zaskrzeczał** myszołów ...' (364)
   Mewing is a high-pitched clear sound, *skrzeczenie* implies interchanging high-pitched and low-pitched notes and a hoarse rather than clear sound.
   'It [a buzzard] **mewed**, majestic, golden, apotheosized, against a dark cloud above.' (456) 'Majestatyczny, złoty, uświęcony, **skrzeczał** spod ciemnej, wysokiej chmury.' (543)
   For the translator, prototypical raptors squawk or croak (*skrzeczeć)* and he betrays inhibitions towards the 'softish' verb 'mew' (*miauczeć, kwilić*), which in actual fact mirrors the buzzard's voice.
b) 'A motionless human figure loomed beside the road, holding out a dead bird with an outstretched arm. Dan glimpsed a flat bill, a spiral of red and green on the head.
   "What was it?"
   "A **teal**."' (621)
   '... Dan zdołał dostrzec płaski dziób oraz czerwono-zielone pióra na głowie.
   Co to było?
   Cyranka.' (732)

Although the translator is given clear clues, he ignores them and instead of writing ***cyraneczka***, puts: ***cyranka***, which does not have a spiral of red and green on its head and which bears a very different name in English – **garganey**.

III. Fowles (1985: 113) – Fowles (1997: 114):

'[T]hen slightly tilts his head, like a **robin** alert to prey' – '... jak **gil** gotów coś upolować'. Sadly, the existence of the passage on page 18 , in which the bullfinch appears and is correctly translated as *gil*, did not alert the translator to being more careful with the word robin. The robin must be ***rudzik***, especially as *gil* is a vegetarian, which is never alert to prey.

IV. Iwaszkiewicz (1969) – Iwaszkiewicz (2002):

'[P]odwórze porośnięte **lebiodą**' (251) – 'a large courtyard overgrown with **chickweed**' (177)
The translator does not seem to realise that *lebioda* is an alternative name of what is scientifically labelled as *komosa biała,* for which Anioł-Kwiatkowska (2003) gives the following English equivalents: **fat-hen**, pigweed, lamb's quarters goosefoot and Blamey, Fitter, and Fitter (2003) has it as 'fat hen'. Another plant sometimes called *lebioda* is *lebiodka pospolita* – the wild marjoram. The most likely plant overgrowing the courtyard is the fat hen. 'Chickweed' stands for a number of completely different plants. According to Anioł-Kwiatkowska (2003), without any qualifier, it refers to *gwiazdnica gajowa* or *gwiazdnica pospolita.* With various adjectives in front, it denotes five other species called *gwiazdnica*, for instance, *gwiazdnica zaniedbana* is 'great chickweed', one species called *mokrzycznik* and four named *rogownica.*

V. Kłosowscy (2003):

Olsy – **bog alderwoods** (20, 50, 141)
Bagienny ols – **bog alder forest** (50)
Alders do not grow on bogs, a habitat too poor in nutrients for them. *Ols* is 'swamp alder forest'.

VI. Kłosowscy (2005):

a) 'Oto majowy poranek na Bagnie Bubnów, jednym z ocalonych tu przed melioracją, rozległych **torfowisk**.' – 'A May morning on the **marshes** called Bagno Bubnów, one of the vast local **peat bogs** that escaped melioration' (98). Bagno Bubnów is certainly a **fen**, not a bog.
b) 'Sławne są zwłaszcza niewielkie, leśne jeziorka, tzw. suchary wigierskie, zarastające stopniowo od brzegów kożuchem mchów.'

– 'It is particularly known for its small forest **ponds**, the so-called *suchary wigierskie*, gradually overgrowing with a carpet of mosses.' (153). *Suchary wigierskie* are natural lakes. 'Ponds' suggests that artificial digging may have taken place. The insertion of 'natural' before 'small forest pond' would eliminate a downgrading ambivalence.

c) The Polish ***torfowisko*** is rendered as **peat bogs** in the following fragment of the English translation: 'The marsh fleawort gets its Polish name (*starzec*, old man) from the fruitification that resembles grey beards. This yellow-flowered, rather rare plant adorns in full summer the moist meadows and **peat bogs**, ...' (174).
Although *starzec błotny* has four Latin names, it is not difficult to find its correct English equivalent. Once found, for example in Moor (1984), it becomes apparent that the plant occurs in marshes and fens, **not in bogs**. Again, the incorrect translation of the Polish *torfowisko* (peatland, in general) has far-reaching consequences. Simply, the marsh fleawort's chances of survival in nutrient-poor bogs are slim or none. This error is also classed above as a wrong dictionary equivalent.

d) 'gody **jelenia**, zwane rykowiskiem' – the **deer** rut (253)
'**łanie**, samice **jelenia**' – '**does**, female **deer**' (319). Doe is a female of roe and fallow deer. *Jeleń* refers to the red deer, whose female is the hind. Note that the same errors taken from other sources have been classified as misread syntagms. As they can stem from insufficient knowledge, their representation from just one publication appears here.

VII. Kobyliński (1984: 1):

[Czarna Wisła] bije z **trzęsawiska torfowego –** it springs from **a peat bog. Bogs** are not fed by groundwater. Areas with springs are usually **fens.**

VIII. Kossak (2001):

a) '[rzeka] tworzy zakola i starorzecza' – 'forming bends and **backwaters**' (45)
'szeroko rozlane starorzecze Narewki' – 'the river's extensive **backwaters**' (207)
'Backwater' is a body of water that was created by a flood or tide, or by being held or forced back by a dam. In Polish, it corresponds

to *cofka*, *wody zalewowe*, *zalew*. *Starorzecze* is a former river bend naturally or artificially cut off from it, forming a eutrophic lake, turning into dry land at a relatively fast rate – in English ‘ox-bow lake’.

b) Female elks breed their calves ‘on inaccessible elevations **grown with alders**’ (161). Anyone who knows a little about nature would consider this an extremely unusual situation. Alders grow in low-lying wet areas. In the Polish original, one is not offered the same exciting news, though. ‘[N]a niedostępnych wyniesieniach wśród olsów’ means simply; ‘on inaccessible elevations **among alder swamp forests**’.
c) ‘piętnastocetnarowe [byki żubra]’ – ‘15-hundred-kilogramme [bulls of European bison]’ (240). The calculation is wrong since 1 cetnar = 50kg, 15x50kg = 750 kg. In the light of the fact that the bulls weigh up to 920 kg, but usually far less, the miscalculation spreads serious misinformation.
d) Fallow deer are ‘**nieprzebierające w pokarmie’** (323), which means that they are not choosy as far as their food is concerned. The translator makes them ‘**omnivorous**’, which puts them together with such animals as raccoons or pigs. The truth is that fallow deer are herbivorous animals that eat a wide range of plants.
e) ‘Kogut rozkłada i składa wspaniały **wachlarz ogona** ...’ – ‘The cock [of a capercaillie] furled and unfurled its beautiful tail **like a lyre**’ (474) – Lyre-shaped tails are characteristic of black grouse, not capercaillies. ‘Like a lyre’ is an unnecessary addition of the ignorant translator. It can be classed as a misinterpreted scene, as done in 4.2.2.2. V. above.

IX. Miłosz (1989) – Miłosz (1981):

a) ‘**tracza**, szarego z rudą głową i z haczykiem na końcu dzioba’ (181) – ‘one **merganser**, gray, with a ruddy head and a hooked bill’ (200)
   The description points to the **goosander** rather than merganser, and additionally, the former is usually called ***tracz nurogęś***, whereas the latter ***szlachar*** without ‘tracz’, although both belong to the group of *tracze*. The translator is apparently unaware of the clues presented above.
b) ***pardwy*** (216) – **ptarmigan** (238)
   The birds feature in a description of a hunt in the bogs in Lithuania. For the hunters it was obvious what species it was, because none

other lives there – ***pardwa mszarna*** (**willow grouse**). There is also ***pardwa górska*** – **ptarmigan**, but it has a different range of distribution. The translator must have assumed that *pardwa* was *pardwa*, no matter the fine distinctions between them. Hence his random and irresponsible choice. He even disregarded such helpful clues as the description of the plumage: 'Pardwa była **biało**-brązowa z nogami w getrach, śnieg skrzydeł odcinał się od reszty ciała', which he did not render with due accuracy: 'The ptarmigan was a **beige**-brown bird in spats, its snow-white wings in stark contrast to the rest of its body.' The bird was ***biało***-*brązowy* – **white**-brown. The ptarmigan is marked in shades of grey rather than brown. Thus, not only the place of occurrence but also the plumage, clear indicators of the species, did not suffice to prevent the translator from making a serious error, especially as the context enhances the main character's (Thomas) satisfaction at being able to add a new species to his record. The precise name did matter to Thomas. And so it should to the translator.

X. Sidło, Błaszkowska and Chylarecki (eds) (2004):

a) **lasy łęgowe** (499) – **alder carrs** (501)
b) 'Ostoja obejmuje duże **torfowisko niskie** o nazwie **Bagno** Wizna' (263) – 'The IBA includes a large **raised peatbog** called the Wizna **Swamps**.' (266). The first problem is that raised peatbog means *torfowisko wysokie*. This mistranslation is particularly perplexing because two lines later we find the following sentence: '**Low peatbogs** cover ca. 70% of the IBA'. Although the correct term is 'fens', anyone can see that a different type is meant – low rather than raised. The second problem concerns the word used in the name of the IBA. **Swamps** are usually permanently underwater. That stands in stark contrast with the information that the area was completely drained for farmland use, mainly as hay meadows and is only regularly flooded by the Narew river in early spring. If the name must be translated at all, 'marshes' seems the best choice since it is traditionally given to such famous and very similar places, as the Biebrza Marshes and Prypet Marshes[67]. The same objections should be raised against translating Bagno Całowanie as The Całowanie **Swamp** (only a small part of it is swampy) and

[67] See also definitions in Blamey, Fitter, and Fitter (2003) and the *Book of the British countryside* (1981).

calling this **fenland** 'one of the largest **peatbogs** …'. In the same book (477), another valuable fen, Bagno Bubnów, is named, inconsistently but more realistically, Bubnów **Marsh.** However, the description that follows is again inadequate since Bubnów Marsh is defined as **'a wet peatbog'** (478). A particularly striking clash occurs between the image generated by the Polish ***torfowiska węglanowe*** and its translation – **calcareous marshes** (486, 514) or **peatbogs** over calcareous soil (486). Marshes are not on peat. Peatbogs in turn are far too nutrient-poor to look as lush as a **calcareous fen**. Sometimes, though, the translations are correct despite the general mess, for example, **Torfowiska** Orawsko-Nowotarskie are Orawa-Nowy Targ **Bogs.** It is indeed true that they are a complex of raised bogs not marshes, swamps or fens.

c) ***dęby szypułkowe*** (363) – **White Oaks** (365). The wrong name appears in the description of the largest European concentration of ca. 500 year-old **pedunculate oaks** at Rogalin. The place is well-known in Poland and it comes as a surprise that an alien (North American) species has been growing there for so long. They would have had to be planted by Christopher Columbus.

XI. Sawicki (2003b):

a) 'Są one niezbędne dla niektórych **dziuplaków**, które nie potrafią wykuć dziupli w żywym, twardym drewnie, jak np. sikory: czarnogłowa i uboga.' (64) – 'They [dead trees] are very much needed by certain **hollow drilling birds** such as the willow tit and marsh tit, which are unable to drill hole in the living hard wood' (65). *Dziuplaki* are **hole nesters** – birds which nest in holes, not only in trees. Some of them excavate their own holes in hard living wood, some in dead wood but some just occupy natural holes or those made by other species. Categorising *dziuplaki* as hollow drilling birds in the context above wrongly narrows down the number of possible members and makes readers wonder why creatures that can drill cannot do it in any type of wood. The point is that some of them cannot drill but they remove soft dead matter by scooping out a hole.

b) 'eutrofizacja … prowadzi do zbyt szybkiego starzenia się jezior i przekształcania ich w **torfowiska**.' (70) – 'eutrophication ... leads to the faster aging of the lakes, and to their premature change to **marshes**.' (71)

In the description of this process it is necessary to be precise and explicitly write about peatlands rather than marshes.

c) **bór bagienny** (76) – **swampy forest** (77)
It is not any type of swampy forest. Here again the folk understanding that *bór* means just 'forest' sneaks into a specialist context. See also *pomorski bór mieszany* – Pomeranian mixed forest.

XII. Smartwood (2004: 7):

Bog alder forests – alder forests do not grow in bogs because of the low fertility of this habitat. They require fen peaty substrate, much more nutrient-rich (see I. above).

XIII. Sobociński (2004: 131):

'w Narwi' – 'in Narwia'. This strange form appears despite the fact that the nominative of the name 'Narew' is used earlier in the text and is self-evident for everyone acquainted with the geography of the Białowieża Forest.

4.2.4. Meta-translation errors

According to Hejwowski, such errors result from the flouting (often due to ignorance) of commonly accepted translation rules.

4.2.4.1. The wrong choice of translation technique

The problem concerns the decision about 'what should be translated and what should be transferred or replaced by a functional equivalent' (Hejwowski 2004a: 221). Our database shows that we ought to focus on titles of publications and names of organisations, institutions and forms of nature protection and management. Admittedly, translation theories treat titles and names as a special case, which I do not call into question. I have singled out only those examples in which the following conditions are met simultaneously: 1) nature terminology is used 2) basic rules of translation must be observed 3) poetic licence and marketing considerations cannot take priority over 2). The unjustified popularity of hypernymy also draws my attention.

4.2.4.1.1. Translating titles and names

I. Sawicki (2003a: 58):

> It would be no use to look for this species [Mediterranean shrike] in most books on European avifauna published by now. In the Lars Jonsson guide (issued in 1994) it is listed as a subspecies of the great gray shrike (*Lanius excubitor meridionalis*). It was first described as a separate species in early nineties ... In the **Atlas of the Hatching Birds of Europe EBCC** it is still called the southern shrike and its new valid name – Mediterranean shrike – was finally established in 1999 by the nomenclature of the birds of the world created by the specialists of the Polish Ornithological Section of the Faunistic Society. (59)

The translated text presents two problems with imagery and accuracy of the message (all other errors will not be discussed). Firstly, the translator has rendered the Polish translation of the title (*Atlas ptaków lęgowych Europy EBCC*) of the book, published in English, back into English. Hence, instead of '**Breeding Birds**' we receive '**Hatching Birds**', which is the phrase frequently used by the translator but it is quite awkward. In this context it simply gives the wrong referent, evoking unnecessary processing effort – to use Relevance Theory terminology, especially because the real title of the original appears in the bibliography at the end (*The EBCC Atlas of European breeding birds: Their distribution and abundance*). Secondly, in 1999 the Polish Ornithological Section of the Faunistic Society established the Polish name of the species in question – ***dzierzba śródziemnomorska*** – the English name must have been introduced by somebody else, perhaps at a different time. Certainly, no Polish organisation can impose standards concerning English nomenclature. In the light of the remarks above, the name 'southern shrike' does not seem credible. It is just a literal translation of the Polish ***srokosz południowy***. Without access to the atlas quoted, disappointed with the illogicalities noticed, one has the right to approach the rest of the text with mistrust. Other sources include *Lanius meridionalis,* with its English name being the southern grey shrike.

II. Sawicki (2003b):

> a) Obecnie ma on się stać **Obszarem Specjalnej Ochrony (OSO)** – zgodnie z założeniami Dyrektywy Ptasiej, a być może w przyszłości, spełniając kryteria **Dyrektywy Siedliskowej**, Specjalnym Obszarem Ochrony, chronionym na mocy odpowiedniej ustawy dotyczącej Europejskiej sieci ostoi Natura 2000. (26)

> Currently, this area is proposed to be classified as an **Area of Special Protection (ASP)** – according to the Bird Directive, and Sites of Community Importance (SCI) – according to the **Bird Directive** – protected by law as a part of European network of Important Bird Areas within the **Nature 2000**. (27)

First of all, the name of the European Union's programme is **Natura 2000** and looks the same in any language. Within this programme **Special Protection Areas (SPAs)** are established on the basis of the Bird Directive and Sites of Community Importance (SCIs) on the basis of the Habitats Directive.

b) *Lasy Państwowe* (68) – **National Forests** (69)
In Britain the term 'National Forest' brings totally different associations. Since the 1990s the British have been trying to plant woods on both public and private spaces in the Midlands that would collectively form the National Forest. The project's aim is to increase the forest cover in the UK because it is extremely small. The name of the institution which manages the **state forests** is the Forestry Authority. In Poland most forest cover is managed by the State Forestry Authority, so they are state forests.

III. Sobociński (2004: 143):
The author is said to be 'the editor of the Polish Swamps Web site' 'on behalf of the Protect the Swamps Association'. An organisation bearing such a name has never existed. The officially accepted name it used before adopting a different one: Centre for Wetland Conservation was **'Save Wetlands Association'**, which stands for 'Stowarzyszenie Chrońmy Mokradła'. It exists to save wetlands, not solely swamps.

4.2.4.1.2. Using hypernyms

I. Durrell (1973): – Durrell (2001):

'[A]t others [the galago's ears] standing up pricked and straight like anaemic **arum lilies**' (106) – 'kiedy indziej zaś sterczały wyprostowane jak **blade lilie**' (77)
'Arum lily' is ***kalia*** or ***białokrasa/belikrasa*** in Polish. Its flower sticks out in a very characteristic way, very different from popular garden lilies or wild martagon lilies. The English version brings to mind a concrete unique image. The Polish text includes a vague reference to 'pale-coloured' lilies.

If someone can imagine the petal of a white water lily (*grzybienie białe*), they might not be too wide of the mark. But any other lily does not have blooms even distantly resembling the arum lily's. In this situation Durrell's comparison does not help visualise the galago's ears and seems rather artificial. Not many readers would blame the translator, the author quite unjustly bearing the brunt of the criticism. Incidentally, 'arum lily' appears in *Birds, beasts and relatives* by Durrell (62), where it is rendered correctly as *kalia* (88).

II. Durrell (1976: 53) – Durrell (1999: 80):

'[T]he clear blue of a hedge-sparrow's egg' – '… jajka płochacza'
'Hedge sparrow' or 'hedge accentor' is ***pokrzywnica***, formerly known as ***płochacz pokrzywnica***. But there is another ***płochacz*** – ***halny*** (*Alpine accentor*) and so it happens that both have similar eggs. Translating it as '*płochacz*' is insufficient and misleading.

III. Fowles (1975: 242) – Fowles (1994: 260):

> "Oh Kate, look! There are my butterfly orchids". And Annabel leads the way up a little sunlit clearing in the trees beside the path, to where five or six slender, white columns of the delicate flower stand from the grass ... . "Why are they yours?" " Because I found them last year." ... "They're wild orchids", says Annabel. "Butterfly orchids".

At that the other participants of the outing come and '[t]hey seem disappointed, a little at a loss when they see how small and insignificant the plants are.' In the Polish version Annabel exclaims: 'Kate, patrz! Moje dzikie **orchidee**!' This sounds like an utterance made by an inspired but ignorant artist, whereas Annabel was very down-to-earth and matter-of-fact about her find: 'Kate, patrz! To są moje **podkolany białe**!' (translation mine). Admittingly, the proposed Polish version sounds much more scientific because the English name of the flower allows people to categorise it within the well-known group of orchids, whereas in Polish it does not include the group's name, which is ***storczyki***. Therefore, in order to make the text sound 'matter-of-fact' but not coldly scientific the following can serve as a compromise: 'Kate, patrz! To są moje **storczyki**, **podkolany białe**!' The word *orchidee*, implying some big impressive and exotic flowers, appears to be particularly awkward in the context of the disappointment described above.

IV. Fowles (1977) – Fowles (1982):

a) 'What is it? [Conchis:] Otus scops. **The scops owl**. It is very small. Not twenty centimetres.' (105) – '... Otus scops. **Sówka**. Jest maleńka ...' (125). In such a context, the translator is expected to find a native name for the species – in Polish ***syczek***, firstly in order to explain the Latin and secondly, because the speaker claims to be a specialist in birds. What the reader receives in Polish is a non-scientific diminutive whose function seems to be the strengthening of the words describing the creature as very small. Later, the bird is again treated as a non-specific owl: '**The scops owl** called from somewhere nearby.' (359) – 'Gdzieś obok odezwała się **sowa**' (429). This error may stem from the insufficient knowledge of the subject-matter and be regarded as a wrong translation strategy (overdomestication).
b) 'The gulley was finally negotiable, though it was a tough scramble up the far side through some **disagreeably sharp-thorned smilax**. Once through that I was able to run again. The carob came into sight below.' (142)
'Wreszcie udało mi się przebyć parów, choć z tamtej strony trudno mi było się wdrapywać przez **niemile kolczaste zarośla**. Potem znów puściłem się pędem. Zobaczyłem świętojańskie drzewo.' (170)
The Polish translation, as in many other places, lacks specificity of description. Instead of **'niemile kolczaste zarośla'** a lot more Fowles-style would be: **'kłujące krzaki kolcorośli'** because 'smilax' is a plant called ***kolcorośl*** in Polish.

V. Fowles (1978) – Fowles (2000):

a) 'another of his secret flowers, the honeycomb-scented orchid *Spiranthes spiralis,* blooms on the old meadows' (9) – 'inn**y** z jego sekretnych kwiatów, mał**e** pachnąc**e** miodem orchid**ee**, które właśnie teraz kwitną na starych łąkach' (16). Apart from the lack of number agreement – it should be ***inne*** to agree with *orchidee*, or if we want to retain *inny,* the plural has to be turned into the singular 'mał**a** pachnąc**a** miodem orchide**a**' – there are more serious problems here. Details concerning nature are very important for Fowles, and for the main character. That is why a concrete name of the species is mentioned. In the Polish text no species name is given. Instead, Daniel is said to look for 'orchidee', flowers coming from the tropics, prototypically quite sizeable. Certainly, since the flowers in question are such a secret, the mysterious Latin name can be preserved unaccompanied by the Polish equivalent. Alternatively, the Polish name ***kręczynka jesienna***, as it sounds equally mysterious to most Poles, might appear. Writing *storczyk* not *orchidea*, though rather general, is also tolerable. But because the orchid is very small, delicate and grows in temperate climates, one cannot suggest it is a big, exotic flower.

b) **chipmunks** (344) – ***wiewiórki*** (413) (but on page 421 they are no longer hypernymous ***wiewiórki ziemne***, which is true in reference to the west of the United States of America, the eastern chipmunk being *pręgowiec amerykański*).

c) 'The first spurt of the milk on the zinc. Like the **woodlark**.' (375) – 'Odgłos pierwszego strumienia mleka o blachę. Jak skowronek.' (449) Although the woodlark is *skowronek*, it is *skowronek borowy*, whose up-to-date scientific name is ***lerka***. In this particular case it does matter which species is meant, because the skylark does not sound like the first spurt of milk on the zinc of a pail.

d) a **tawny owl** (426) – ***sowa*** (509) instead of ***puszczyk***.
For some reason known only to himself, the author wants to be precise, while the translator does not seem to feel the same.

VI. Fowles (1985) – Fowles (1997):

a) 'the sound of a **missel-thrush**, from far across the valley, barely audible, as fragmented as the muffled girl's voice.' (12) – 'trele **drozda** …' (13) (see the comment for VII c immediately below)

b) 'A. I there came upon Jones, who was fishing for **roach** in the stream,…' (152) – 'Natknąłem się na Jonesa, który łowił **ryby** w strumieniu ...' (143)
Not a serious trasgression, it is yet another example of lack of respect for the author's love of the particular rather than the general. Why cannot it be 'który łowił **płotki** …'?

c) 'And I saw there was now no further of trees, be it not for some few twisted **thorns**, but sward and **bracken** above, rough leaze, that ran to somewhat of a basin, ...' (222) – '… tylko trochę karłowatych **kolczastych krzewów** a wyżej murawa i wysoka **paproć**, potem skalisty kawałek, dochodzący jakby do niecki z wodą.' (211)
Being imprecise in this fragment is justified in the case of 'thorns', which could signify hawthorns or gorse, so the author himself is

rather ambiguous. As regards bracken, writing 'paproć orlica' would not change much but would be more Folwesian.

VII. Fowles (1996) – Fowles (2003):

a) 'its **bracken** that grows seven, eight feet tall' (71) – '**paprocie** dochodzące do trzech metrów wysokości' (65)
   again **bracken** (100) – ***paprocie*** (91)
   'through the dead bracken and living gorse' (252) – 'przez zwiędłe **paprocie** i kwitnący janowiec' (235)
   The translator hardly ever gives the name of this particular species – ***orlica***. In the one place it is actually given – 'in the gorse and **bracken**' (249) – 'w zaroślach janowca i **orlicy**' (233) - it does not seem to have any special function in the context, a function substantially different from the other contexts quoted. Most often it is just *paproć*, which is not wrong, but as a type of fern that has played a very important role in the economic and natural history of some regions of Great Britain the bracken deserves a more special treatment. Bracken was used to thatch roofs and nowadays is famous for its omnipresence due to overgrowing large tracts of land and, in effect, choking the growth of rarer and more valuable plant communities.
b) 'It had brought out swarms of spring butterflies, those **brimstones**, **orange-tips** and **green-veined whites** ...' (160) – 'Przywitał roje motyli wiosennych, siarkowożółtych z pomarańczowym obrzeżeniem i zielonymi żyłkami' (150). This can be interpreted as the description of a swarm of individuals representing one species so fancifully coloured. In fact, the names of three different species are enumerated, whose Polish equivalents are (in the plural): ***latolistki cytrynki, zorzynki rzeżuchowce*** and ***bielinki bytomkowce***. The employment of the hypernym *motyle*, accompanied by their description, may result from the translator's insufficient knowledge or from adopting the wrong strategy of simplifying everything overspecific not to distract readers from the romantic story, already complicated enough.
c) 'In one of the great ash trees below hidden **missel-thrush** was singing, wild-voiced beneath the air's blue peace.' (168) – 'Na jednym z wielkich jesionów rosnących niżej śpiewał **drozd** – czysty, swobodny głos rozbrzmiewał wśród ciszy.' (158) The comment here is similar to that concerning the herring-gull in e) below. There are many species of thrushes (***drozdy***). For some

reason Fowles singles out the mistle thrush (current spelling), which is ***paszkot*** in Polish, the translator prefers to be vague and general for a rather obscure reason.

d) 'But she stared at the **hart's tongue ferns** at her feet' (180) – 'Ale Sara wpatrywała się w kępę **paproci** u jej stóp' (169)
The hart's tongue fern is ***języcznik zwyczajny*** in Polish and is a fern that differs a lot from what most people imagine when they hear the word 'fern'. In Poland it is very rare, in Britain quite common. Again, it is difficult to understand why Fowles' vision of the world, so sensitive to detail has been simplified and flattened.

e) 'Two **herring-gulls** flew over, raucously' (240) – 'Dwie **mewy** przeleciały mu nad głową, krzycząc chrapliwie' (224)
For the author it is important to notice that the two gulls are herring-gulls. Ignoring this the translator retains gull (*mewa*) but makes it an unspecific one, not ***mewa srebrzysta***. Thus, instead of the big, light-coloured gull, we can just as well imagine the small black-headed gull (*śmieszka*), which is quite common in Poland.

VIII. Hryniewiecki and Okołów (2004: 114):

The following quotation illustrates a range of mistranslation issues, one of them being the employment of a hypernym 'willows' instead of a lower-rank term 'grey willow/grey sallow': 'Wśród **łozy, wikliny**, rokity' – 'Among willows, osiers and creeping willows'. There is some mess in nomenclature. *Łoza* is the common name for *wierzba szara* (Czekierda 1992, Anioł-Kwiatkowska 2003), *wiklina* is *wierzba purpurowa* (ibidem) but *wierzba szara* is also known by the name of *wiklina* in places (Anioł-Kwiatkowska 2003). Czekierda (1992) gives the following English equivalents of *wierzba purpurowa (wiklina)*: purple willow, red willow, purple osier, red osier; Anioł-Kwiatkowska (2003) lists: purple osier willow and purple willow. W*ierzba szara (łoza)* in both dictionaries is grey willow or common sallow. Blamey, Fitter, and Fitter (2003), in turn, gives for this species the names grey willow, grey sallow, listing common sallow under goat willow, which is *wierzba iwa* in Polish. The name osier in Blamey, Fitter, and Fitter (2003) is given to *salix viminalis*, which is in Polish *wierzba wiciowa* aka *wikwa* (both in Czekierda (1992) and Anioł-Kwiatkowska (2003) as common osier). As far as *wierzba rokita (salix rosmarinifolia)* is concerned, in Czekierda (1992), it is rosemary-leaved willow but in Anioł-Kwiatkowska (2003) – creeping willow. Blamey, Fitter, and Fitter (2003) informs us that creeping willow is *salix repens* (*wierzba płożąca* according to the other two sources). To make matters even more complicated, Anioł-Kwiatkowska (2003) translates *wierzba płożąca piaskowa* as sandy or sand willow, Blamey, Fitter, and Fitter (2003), which includes only the species growing in the wild in Britain and Ireland, does not mention *salix rosmarinifolia.*

In this situation we have two possibilities: either to be scientifically correct and write 'Among **grey, purple and rosemary-leaved willows'** or use broader categories whose boundaries do not seem strictly determined in the source and target languages alike: **'**Among **sallows, osiers and rosemary-leaved willows'.** But by no means should we resort to as high-rank a taxon as the genus 'willows', which contains about 400 species.

IX. Iwaszkiewicz (1987) – Iwaszkiewicz (2002):

a) '... zarośnięta **tatarakiem** i rzęsą struga, kryjówka kaczek i zimorodków.' (410) – '... ran a stream overgrown with **rushes** and duckweed, where ducks and kingfishers lay in hiding' (24). *Tatarak* is 'sweet flag' and may only be part of 'rushes'. Strangely, the translator seems to remember that 38 pages later in the book.
b) 'Pamięta, jak kupowała **jagody** od baby, jak zsypywała z dzbana górę poziomek na półmisek, ...' (366) – 'He remembered her buying **berries** from an old woman and pouring a mound of wild strawberries from a jug into a dish' (132)
and:
'zapach siana i **jagód**' (458) – 'ripe and pungent with hay and **berries**' (85)
It must be pointed out that wild strawberries are also 'berries'. In the English sentence, what was being bought could range from bilberries to strawberries and gooseberries (see X. below). As the Polish *jagody* refers in many regions to bilberries, however incorrect it is from the scientist's perspective[68], the image evoked in most Polish minds is very specific, by which it contrasts sharply with the extremely schematic one in the English version.

X. Kossak (2001):

a) In the English edition in the alder swamp forest on hummocks at tree bases one can find e.g. '**berries**'. In the Polish one we have ***borówki*** (30). Both versions are ambiguous, but the translator,

[68] See Chapter Three under the heading *Borówki/Jagody.*

rather than filling in Ingarden's places of indeterminacy[69] has chosen to enlarge them. The choice of '-berry' words in English is impressive. Fortunately, the clues concerning the habitat narrow down the list of most likely suspects to the crowberry, cranberry and wild strawberry. Their Polish equivalents are, respectively, *bażyna*, *żurawina* and *poziomka*, none of which is traditionally categorised as *borówka*. On page 32, *Na pobliskim* ***borówczysku*** is interpreted as 'In a **cowberry field** close by'. This time the reverse happens – a range of possibilities is narrowed down and fits the context well (the capercaillie hen is walking her young there). Similarly correct is the rendering of ***jagodziska*** in: 'rich in brushwood and undergrowth, especially **bilberry**'. *Jagodziska* is not a scientific term and is used in many regions in Poland to denote areas covered by bilberry. Scientifically *jagody* are 'berries', encompassing, like in English, a great variety of plants but, unlike in English, rarely being included in their names.

b) 'Until recently they [hunters] killed **grouse**, hazel grouse and ruffs during their mating time.' (478) – A British reader is likely to imagine red grouse being hunted in Poland, because 'grouse' is how they casually refer to red grouse, a common game bird in the UK. In the Polish text, however, ***cietrzew*** (black grouse) features, red grouse being totally absent from our country.

XI. Kłosowscy (2005: 89):

'[Z]ieleń **jagodników**' – 'green clusters of **berries**'
The accompanying picture shows **bilberries** carpeting the slopes of the Tatra mountains.

XII. Lewandowska and Lewandowski (1994):

a) *mewa śmieszka* – **a gull** (28)
Unlike in Polish, no specific information is given in English, which is expected because the name accompanies a photograph. Everyone can see a gull there. Not many people know which species. In order to bring the caption to the same level of specificity '**black-headed gull**' should be given.

b) 'larwa **zawisaka gładysza'** – 'a **sphingid** larva' (47)

---

[69] The term was coined by Roman Ingarden in his *O poznawaniu dzieła literackiego*. 1937. Lwów: Ossolineum, translated into English in 1973. In the context of translation work it was brought to my attention by Tabakowska (1999: 220).

Although the authors do not help the translator by giving a not-fully-correct name and no Latin name, writing 'sphingid' determines the whole family of moths, *Springidae* (***zawisaki***), rather than a concrete species, which is **the elephant hawk-moth** – ***zmrocznik gładysz*** *(Deilephila elpenor)*. As such it is not sufficient.

c) *kraśnik siedmioplamek* – **zygaenid** (49)
No wonder the translator had problems since a species with such a name does not exist. The picture next to the caption shows ***kraśnik sześcioplamek*** – **six-spot burnet**, which is one of fourteen zygaenids occurring in Poland. The incorrect Polish name perplexed the translator to the point that s/he chose to put down only the name of the whole family rather than trying to guess whether there was an error or insufficient data at his/her disposal. Without Latin it is not easy to look for English equivalents in such situations. Nevertheless, the appearance of the hypernym instead of the specific name is unwelcome.

d) *strojnica baldaszkowata* – a **pentatomid** (81)
Again the authors make the translator's task more difficult than necessary, making an error in the name, which is ***strojnica baldaszkówka.*** The Latin name is not provided, so finding takes some time. It turns out that there is no English name and it appears in guide books as ***Graphasoma italica*** or ***lineatum***. Pentatomid only tells us that the bug belongs to the superfamily of *Pentatomoidea* without even narrowing down one's choice to a particular family.

XIII. Lewandowska and Lewandowski (1999):

a) Rezerwat „Bocianowskie Bagno" utworzono w celu ochrony **kompleksu zróżnicowanych leśnych zbiorowisk borowych** porastających tereny wydmowe i obniżenia międzywydmowe oraz towarzyszące im mszary bagienne z jeziorkami torfowiskowymi.' (16) – 'The Bocianowskie Reserve – to take special care of **group of plants** growing in dunes and depressions between them, morasses and swampy peat lakes;' (20)
In Polish we read that special protection is given to **'a complex of diverse coniferous woodland communities'**, in English we have a mere 'group of plants'.

b) ***jaskier ostry*** (17) – **buttercup** (21).
There are many buttercup species, from the numerous *jaskier bulwkowy* (bulb buttercup) to *jaskier szczeciniasty* (bristly

buttercup), which does not occur in Poland. Anioł-Kwiatkowska (2003) lists forty six in all, twenty five of which grow in Poland. This one is the **meadow buttercup.**

c) *listera jajowata* (18) – twayblade (21)
There are two species of twayblades, **the common twayblade** being ***listera jajowata*** and **the lesser twayblade** – ***listera sercowata.***

XIV. Miłosz (1989) – Miłosz (1981):

a) 'Stromy brzeg, ... podziurawiony przez **jaskółki brzegówki**' (76) – 'The steep bank, bright with yellow clay, was riddled with **swallow's** nests' (83)
Swallows do not dig holes in river banks. Such nesting habits are characteristic of **sand martins**. Swallows and martins (sand and house) are sometimes collectively referred to as swallows. In the context of nest holes in the steep bank the use of this hypernym is strikingly inappropriate. Swallows and house martins build their nests on and in buildings – a rather familiar sight in villages and small towns.
b) '[puchacze] rozszarpywali **sarenki** i zające' (95) – '[eagle owls] once preyed on **deer** and hare' (104)
Attributing to eagle owls enough strength to kill any deer, red deer included, is a gross exaggeration, especially as even roe deer must be either very young or still small and extremely weak to become an eagle owl's prey.
c) e.g. 160 – ***puchacz***, which appears several times in the text, is invariably referred to as '**owl**', not 'eagle owl' (176).

XV. Nowiński (2005: 42):

Olsy **– carrs**
The English term covers both alder swamp forests (*olsy*) and other fen forests, in Polish called *łęgi*.

XVI. Nycz (ed.) (2004: 36, 37, 38):

Bojowniki **– sandpipers**
The picture shows **ruffs**. Their official Polish name is ***bataliony***, *bojowniki* being sometimes used. Sandpipers is too general a category to apply to a concrete species. The ruff belongs to the ***Scolopacidae*** family,

sandpipers in English, ***bekasowate*** in Polish. A number of species which belong to this family bear the name of sandpiper. Some of them are ***brodźce***, some others ***biegusy*** in Polish. The ruff is closely related to some sandpipers, closer than such members of the same family as curlews (***kuliki***) or snipe (***bekasy***), but definitely does not deserve being stripped of its identity. In fact, it is one of very few sandpipers that non-specialists can easily recognise, at least in the mating season.

XVII. Sawicki (2003a):

Biegusy (88) – **sandpipers** (89)
Not all sandpipers are ***biegusy*** in Polish, some of them are ***brodźce*** (e.g. *brodziec pławny* – marsh sandpiper, *brodziec piskliwy* – common sandpiper, *samotnik* – green sandpiper, *łęczak* – wood sandpiper; the latter two were formerly known as *brodziec samotny* and *brodziec leśny* respectively).

4.2.4.2. Additions

I. Kossak (2001: 420):

The following sentence, not being part of the Polish text, is inserted in the English version: 'Their final report contained many remarks criticizing the massive logging done by Century.' The report in question was submitted by the commission of scientists investigating the possibility of establishing a national park in the Białowieża Forest in January 1922. Earlier in the book we learn that the contract with the firm Century was signed in 1924 and the logging terminated in 1929. Why such a perplexing statement has sneaked into the paragraph is indeed mysterious. If associated with certain omissions (see 4.2.4.3.), this addition can be said to contribute to the changing of the overall tone of the book (see the discussion in 4.2.4.6. below).

II. Miłosz (1989: 161/2) – Miłosz (1981: 178):

'Tomasz pełzał na czworakach i podglądał, zbliżyć udało mu się tylko raz, drugi raz był to cietrzew, który czuszikał siedząc na sosence: ...' – 'Crawling up on all fours, he would spy on them, though only once at close range. Another time he **was startled** by the mating call of a blackcock perched in a young pine. *Choo-shee*!'

In the Polish version there is nothing about Tomasz being startled. Conversely, we find out how proud the boy was to approach the bird so closely without scaring it away. That suggests no element of surprise and fright but rather an expected thrill. The distinction made by the narrator between the two sightings of blackcocks at close range results most probably from the fact that the first time he managed to approach the whole group of displaying males and the second time just one cock outside the arena, preparing to join the group or standing too low in the hierarchy to do so. The lack of knowledge concerning the mating behaviour of blackcocks caused the translator to misinterpret the script.

III. Sawicki (2003b):

a) 'The Iławskie Forests, in addition to various types of forests, have also quite often the beach areas. However these beaches are very rarely in their native and unaltered state.' (23)
   This fragment does not appear in the Polish version. Since it overflows with target language errors it comes under scrutiny in 4.2.3.1. together with the co-text.
b) 'Cyraneczka – … w Polsce bardzo nieliczna, jedynie lokalnie nieco liczniejsza, lęgowa (1300-1700 par).' (46) – 'Teal – ... In Poland extremally scarce, locally a little more numbers (1300-1700 pairs), **but locally can be very abundant**' (47). All the language mistakes are bad enough, but more worryingly, the information in bold does not appear in the Polish version and actually contradicts what is said immediately before. In Relevance Theory terms, the foreign reader's processing effort is unnecessarily increased.
c) 'Nie można wykluczyć gniazdowania włochatki, którą stwierdzono tu zimą.' (50) – 'The remains of the Tengmalm's owl have recently been found in winter.' (51)
   What we read in English departs considerably from the message in Polish, which is: we cannot exclude the possiblity that the Tengmalm's owl nests here. Its presence was once reported in winter. Nothing indicates that the owl was found dead. The addition of 'the remains' might be a correction of the author by the translator towards greater specification or, quite the opposite, a misinterpretation of the scene.

IV. Sobociński (2004: 4 and 5):

‘Do tego skrawka **na wschodnich rubieżach** Polski skurczył się obszar dawnych puszcz europejskich.’ – ‘This is a fragment of **Poland’s eastern marshes** that ancient European forests shrank to.’
In the Polish version nothing is said about marshes but about Poland’s eastern borderland, which is not all just marshes.

4.2.4.3. Omissions

I. Durrell (1975: 30-31) – Durrell (1991: 42):

> ... in the dim recesses of convulvulus-covered tree stumps robin-chats hopped perkily in search of grasshoppers, and looked startlingly like English robins; **a pied crow flew up from the ground ahead and flapped heavily away,** crying a harsh warning

> W mrocznych zagłębieniach obrośniętych powojem pni żwawo podskakiwały rudziki szukające pasikoników – przypominały zupełnie angielskie gile, skrzecząc ochryple i ostrzegawczo.

Robin-chats are ***złotokosy*** not ***rudziki*** and, according to Durrell, they resemble ***rudziki*** not ***gile,*** which in English are **bullfinches.** But apart from this mix-up, the translator serves another. She leaves out the above fragment in bold, retaining ‘crying a harsh warning’. In effect, these are the robin-chats that ‘cry a harsh warning’ rather than pied crows (***kruki srokate***). Although robin-chats, like most birds, are capable of uttering a harsh warning call, they do not, as the Polish text implies, **‘croak hoarsely and warningly’.**

II. Fowles (1977): – Fowles (1982):

a) ‘A primrose-and-black butterfly, **a swallowtail**, hovered over the bougainvillaea round the Priapus arbour, ...’ (146) – ‘Żółto-czarny motyl krążył nad bougainvilleą obrastającą altankę Priapa, ...’ (174)
In the Polish text, the butterfly remains unidentified, although the author tells us it is a swallowtail – ***paź królowej***. Fowles’ sensitivity to detail is again remorselessly blotted out in translation.
b) ‘Datura. **The thorn-apple**. It produces hallucination.’ (233) – ‘Datura. Wywołuje halucynacje.’ (279)
Fowles is careful to give the English name alongside the Latin. The translator fails to do that, although amateur botanists, who use

Polish names and are often unaware of the Latin ones, are greater in number than professionals and thus more likely to form a part of the collective average reader. Most of them know ***bieluń dziędzierzawa***, but probably not many can match it with *datura*.

c) 'rare duck and geese, **the divers**, the wild swans, that abounded in all the inlets and lagoons along the shore' (297) – 'rzadkim gatunkom dzikich kaczek, gęsi i łabędzi, które zamieszkiwały laguny i zalewy' (357).
It is probably due to carelessness that 'divers' – ***nury*** have disappeared in translation, but, unluckily, this is the group of birds most confined to the northern regions under description.

d) 'a cool grove of **ilex**' (569) – 'chłodny gaj' (215, vol.II)
'… and I went down the rock staircase to the path through the **ilexes**' (572) – 'a ja zacząłem schodzić po skalnych stopniach.' (219)
In both fragments the trees are missing in the Polish version. In this way the translator shows disrespect towards the author, who does mention their names. Both ilex in English and its equivalent, ***ostrokrzew kolczasty***, sound exotic. This is so because, in English, the author replaced the more commonly used 'holly' with the rarer 'ilex', while in Polish, because our climate is too harsh for the tree and it is only known to a small group of garden lovers as a shrub. Therefore, omitting the tree's name altogether in the Polish text cannot be easily justified by the wish to domesticate.

III. Fowles (1985) – Fowles (1997)

a) 'In a great wood we call **Cleave** Wood, that stretches to the moor and is more **steep** that ne'er a man may walk in many places, more **cleave** than **combe**.' (73) – 'W wielkim lesie, który my nazywamy Gęstym Borem. Rozciąga się aż po bagniska i rośnie czasem na takich **stromiznach**, że człowiek nie zawsze może tam przejść.' (74)
In the English version the steepness of the place is marked four times: in the name of the wood, by the direct description and, at the end, by specifying what sort of geological formation the place was. The Polish text refers to the steepness only once, in the direct description. The name of the wood is mistranslated and the specification concerning the rocks is left out. In order to preserve the spirit of the original the following translation is suggested:

'W wielkim lesie, który my nazywamy **Lasem w Szczelinie**. Rozciąga się aż po bagniska i rośnie na zboczach takiej **szczeliny**, bardziej niż **wąskiej doliny**, czasem na takich **stromiznach**, że człowiek nie zawsze może tam przejść.'

'Lasem na Rozpadlinie' can substitute for 'Lasem w Szczelinie', as it may be perceived as a more elegant version. Considering that the speaker has a lay, folk perspective, the otherwise incorrect 'Borem ...' cannot be rejected either.

b) 'Q. "This chest had been hidden?"
A. "'Twas thrown **in a goyal of thick bushes**, four hundred paces from the road. …"
Q. "A goyal, what is that?"
A. "**A combelet**, sir. A narrow sunken place" (74).

'A. **W gęste zarośla**, o jakieś 400 kroków od drogi.
Q. Dokładnie, gdzie była wrzucona?
A. W **wąską szczelinę**.' (75-76)

'Goyal', or according to the Webster's Dictionary 'goyle', is a steep, narrow valley and must be a regional word, not understood by all native speakers of English, hence the question of the interrogator: 'A goyal, what is that?'. In the Polish text the question is removed as a consequence of the tolerable but preventable elision of the non-standard word. Only the more readily understood 'combelet' is translated and it is done with the help of the perfectly standard Polish ***szczelina***.

IV. Hryniewiecki and Okołów (2004):

a) Despite the fact that before long lists of species have been diligently translated, seven, not particularly difficult to find, are deleted from page 74.
b) 'Fiiilip, Fiiilip, wylej olej, wylej olej – tak w ludowej interpretacji brzmią ich [czajek] krzykliwe głosy' (123). This folk interpretation of the calls by lapwings has equivalent ones in English. A folk name of the bird is quite suggestive too – pee-wit. Here is an example:
When the lapwing, "peeweet," "peeseweep," "wallop," kept screaming and flying around one, he used to call out:
"Wallopie, wallopie, weet (or weep),
Harry the nest, an rin awa' wee't"
or:

"Peesweet, peesweet,
Herry ma nest an awa wee't"[70]

Quoting them instead of the Polish folk interpretation would make little sense but the literal translation would not do any harm as it sounds faithful to the vocal performance of lapwings and not unlike the creations of the British folk. Thus, we can come up with something like: Phiiilip, Phiiilip, spill oooil, spill oooil.

V. Kossak (2001):

a) 'A humming swarm of wild bees clusters round a bear ...' – ... obsiadły, **wiedzącego gdzie szukać miodu** – "miedwieda" (31). In the omitted fragment the Polish reader receives a little guidance on the etymology of the word 'niedźwiedź' = 'bear', which may be quite surprising for those who do not speak Russian, where the word miedwied', unlike in Polish, has preserved the meaningful chunks 'mied' = honey and 'wied' = one that knows. Foreigners may not be interested in the intricacies of the Polish language but in a saga they are likely to expect more than just plain descriptions of nature. Etymological glosses are treated similarly throughout the book. For instance, on page 227 in 'Pszczoły (bczoły – od brzęczeć, buczeć)', the bracketed information that the Polish word 'bee' comes indirectly from verbs expressing the noise bees produce is dispensed with. Nor is the foreign reader offered an explanation why *bartnik* was also called *bartodziej (barć* = a hollowed-out tree for bees to nest; *dziać* = to scoop up a hole). Also the author's analysis of the Polish word *żubr* for the European bison is abbreviated and Slavic names are not mentioned (245).
b) After 'Today beaver colonies exist on several rivers in the forest' the author's commentary 'Jest to dowód, że człowiek może naprawić część krzywd wyrządzonych przyrodzie' (132) is cut out of the English version. Here is my translation of the missing fragment: 'It proves that a human being can right some of the wrongs done to nature.'
c) Red deer have '[a] **light patch** on the rump around the tail ... called a **'mirror'**. In the Polish version red deer have ' **"lustro" ',** which is **'izabelowatopłowa plama obramowana ciemnokasztanową**

[70] The examples are taken from: www.sacred-texts.com/neu/celt/nes/nes02.htm - 33k accessed 11.04.2007.

sierścią' (200). The **'light patch'** is described in detail as 'isabel-fawn' and 'framed with dark chestnut fur'.

d) Passages on Muravyov's policies are skipped (264), which may not turn out to be very wise because foreign readers know about the tzarist rule in Poland next to nothing and need a more insightful rather than abridged historical background.

e) '... obwiesza radionadajnikami i śledzi ich dalsze poczynania tak długo, aż zestresowane zwierzę zginie. Taki los spotkał w latach 90. **prawie** całą populację najwspanialszego puszczańskiego zwierzęcia – królewskiego rysia.' (332)
'Senders are attached to them and their movements and behaviour are followed from a distance until the distressed animals die. In the 1990s such was the fate of **the whole** population of the magnificent forest animal – the regal lynx.'
The translator has skipped one, but crucial word *prawie* (almost). Fortunately, it is **almost** the whole population that suffered from being burdened with telemetric devices, not the entire one.'

f) Between pages 395-400 several, sometimes long, fragments have vanished from the English text. In the original they add to the description of how determined the Polish authorities were to exploit the Białowieża Forest that even the appalling financial situation of the British logging firm Century did not manage to dissuade the government from signing an unfavourable (both for the economy and the environment) contract with it in 1924. They also juxtapose a drawn-out process of working out plans of sustainable management in the Białowieża Forest by a special commission of foresters and simultaneous hasty preparations to sell the right to clear cut large tracts of primeval forest. The left-out passages do not change the main message that the havoc wreaked on the Białowieża Forest by the activities of Century is embedded in the destructive practices local people, foreign invaders as well as the Polish authorities have all been too happy to carry out. Perhaps by way of compensation, on page 402, some very telling statistics, absent from the original, appear: '8,000 ha by the Century company, 6,500 ha by Polish foresters in 1920-1924 and another 5,500 ha in 1930-1939 [of primeval forest were cut down]'. Although the text loses a little in emphasis, its coherence is preserved, which results in the reader remaining unaware of any manipulation.

VI. Lewandowska and Lewandowski (eds) (1999):

'Występują tu turzyce wysokie jak turzyca sztywna – tworząca wysokie kępy, **turzyca pęcherzykowata**, **turzyca dzióbkowata**, ...' (17) – '... their specific vegetation, which consists of tall sedges growing in clusters (e.g. **Carex stricta**), ...' (21)

Admittingly, it is better to give just one example than to provide wrong equivalents but Latin names of the three sedges are readily available. ***Turzyca sztywna*** is ***Carex elata*** or ***Carex stricta*** (**tufted sedge**), ***turzyca pęcherzykowata*** is ***Carex vesicaria*** (**bladder sedge**), while ***turzyca dzióbkowata Carex rostrata*** (**bottle sedge**). The English names might vary, but as proved, official versions exist.

VII. Miłosz (1989) – Miłosz (1981):

a) 'Którejś zimy ... Tomasz widział nad Issą **gronostaja** czy **łasicę**. ... Ani **łasicy**, ani podziwu, ...'' (24) – 'One winter … Thomas spotted an **ermine** on the Issa. … Then there would have been neither **ermine** nor any sense of wonder. ' (24)
   Why is *łasica* (weasel) left out in both cases one can only speculate.
b) Here are the concluding notes of the description of the capercaillie's mating display:
   'Minuty biegły **i nagle w gęstwinie przed nimi trzepot skrzydeł**. Koniec. Odleciał.' (159) – 'Minutes passed. It was over. It had escaped.' (175)
   The left-out fragment takes away much of the suspense: 'and suddenly in the thicket before them the fluttering of wings'.

VIII. Rysiewicz (2000: 17):

'… ochraniał pilnie a chytrze od wszelkiego bartnego zwierza, łakomej na czerw pszczeli **"żołny"** i natrętnych mrówek.' In the English version the word in bold signifying a bird – the black woodpecker – vanishes. We are thus told that the beeman protected his bees against 'all beasts greedy for brood and against intrusive ants'. Perhaps the black woodpecker was not the only brood eater and the generalisation created by dropping the name of the robber is in all likelihood justified. But the omission may be caused by the fact that *żołna* now refers to another bird 'bee-eater', the woodpecker being called ***dzięcioł czarny***. The point is that bee-eaters do not live in the forest and they catch their prey in flight, showing no interest in the bees' brood. The usage of the old regional (eastern) name might have confused the translator, who, just to be on the safe side, resorted to generalisation.

IX. Sobociński (2004):

a) '… białowieskie grądy, olsy, **łęgi**, **bory**, a także nadrzeczne łąki i bagna naprawdę są ostatnią puszczą.' (6) – '... the alder and oak woods, riverside meadows and swamps really make the ultimate forest.' (17)
   It is completely incomprehensible why the translator has decided to leave out two very important types of woodland (riparian and coniferous forests) in the description of what constitutes the last primeval forest. The author purposefully enumerates four of them to underline a great variety, the most valued feature of the primeval habitat. For the discussion of the other errors see 4.2.2.1., p.207.
b) 'W rejonach **podtopionych przez wodę gruntową** lub w dolinach rzecznych spotykamy olsy i łęgi.' (7) – 'In the **wetlands** and along river valleys we shall find alder carr and riverside carr.' (18)
   Wetlands are areas which can be flooded by surface waters as well as be fed by groundwater. Thus river valleys are also wetlands when inundated, and there is no reason to separate the two. Since the author makes a clear distinction between river floodplains and groundwater-fed wetlands it should be preserved in translation.

4.2.4.4. Adopting the wrong translation strategy

This error is committed by translators who do not realise or do not want to accept that translation principally aims at maximal cross-cultural communication. They feel an urge to domesticate texts to the point that all or almost all culture-bound items are filtered out. Others may, in turn, 'foreignise' with zeal, forgetting that the target language has the necessary expressions in its lexicon. Translations involving nature terminology seem to err on the side of overdomestication. It violates the right to expect foreign texts to communicate something new, uncommon, exotic, strange, unpronounceable or 'mind-boggling'. It deserves a particular condemnation in the case of children's literature, in which fresh, receptive minds are offered processed pulp instead of nutritious food, admittedly, demanding a somewhat prolonged digestion time. One pertinent example comes from Polish translations of *The Secret garden* by Hodgson-Burnett and concerns the name of the main animal protagonist. It has been wrongly translated by

Jadwiga Włodarkiewiczowa as *gil*, corrected by another translator, Anna Staniewska, as *rudzik*. Some, but not all editions of the book based on Włodarkiewiczowa's translation, contain the correct version, too. There exist other translations, and in one of them, by Rafał Dawidowicz, *rudzik* becomes *drozd*. While Włodarkiewiczowa can be forgiven, all later translators cannot. When she worked – the beginning of the 20th century – obtaining the most up-to-date information was not easy and, locally, the robin was known as *gil*. Even the pioneer of Polish zoology, Gabriel Rzączyński (1664-1737), gave it this name (cf. Strutyński 1972: 37-38). Later, however, all authorities on the subject called it *rudzik*, including Władysław Taczanowski, whose taxonomy, published in 1882, has survived till today with little change. Nowadays, access to information is much easier and although classification changes frequently, nothing has been altered in the case of the robin since Taczanowski's times. Remarkably, illustrations in the different Polish versions of the book and the film made on the basis of it show the robin. Although *gil* is more familiar to Polish children, the fear that they are unable to absorb a fresh concept and its name smacks of overprotectiveness. When the text contradicts the pictures, instead of being protected from things difficult to absorb, such as misinformation and confusion, children are exposed to them.

I.

a) Durrell (1975: 177) – Durrell (1991: 266):
'**gorse** bushes' – 'zarośla **janowca**'. The gorse is ***kolcolist zachodni*** and it does not grow in Poland, contrary to the similar looking ***janowiec*** represented by several species. Next to hedgerows, dry-stone walls and heather the gorse is the most characteristic landmark of British upland landscape. That cannot be said about *janowiec* in Poland, but the name sounds familiar, unlike *kolcolist* with its rather scientific flavour. It seems then that the translator has opted for domestication rather than foreignisation. However, in the case of biological taxonomy and terminology such a choice breeds confusion. None of the eleven species of *janowiec* listed in Anioł-Kwiatkowska (2003) is called 'gorse' in English, the most frequent name being '**greenweed**'. Most Poles have no idea what *janowiec* looks like, just as they cannot visualise *kolcolist zachodni.* In addition, ***janowiec miotlasty*** serves as an officially recognised synonym of ***żarnowiec miotlasty*** **(broom)**, also known as ***szczodrzeniec miotlasty***. ***Janowiec skrzydlaty*** **(winged broom)**, in turn, is sometimes referred to as ***szczodrzeniec skrzydlaty***. This

situation has been messy enough even without the consequences of domesticating alien names. The same remarks refer to the following:

b) Fowles (1996) – Fowles (2003):
'behind a thicket of **gorse**' (120) – 'za gęstwiną **janowca**' (112)
'he walked round to the **gorse**' (121) - 'podszedł do zarośli **janowca**' (112)
'The **gorse** was in full bloom, the cadmium-yellow flowers so dense they almost hid the green. The air was full of their honeyed musk.' (121) – '**Janowiec** był w pełni rozkwitu, okryty kwieciem tak gęsto, że prawie nie było widać zieleni. W powietrzu unosiła się piżmowo-miodowa woń.' (113)
'in the **gorse** and bracken' (249) – 'w zaroślach **janowca** i orlicy' (233)
'through the dead bracken and living **gorse**' (252) – 'przez zwiędłe paprocie i kwitnący **janowiec**' (235)
The plant so often mentioned deserves a more accurate translation – ***kolcolist zachodni*** not *janowiec*.
c) Fowles (1978: 387) – Fowles (2000: 464):
gorse – ***janowiec***

II. Durrell (1975b: 229) – Durrell (1999: 244)

'the size of **pimpernel** flowers' – 'wielkości biedronek'
Undoubtedly, the translator swapped a little-known flower (***kurzyślad***) for a common insect (ladybird) on purpose. She must have valued the clarity of the comparison higher than faithfulness to the original. On the other hand, if Durrell had wanted to compare something to the ladybird rather than the pimpernel, equally[71] obscure to the British and to Poles, he would have done so. 'Domesticating' the originally 'undomesticated' message may be treated as a misplaced zeal.

III. Durrell (1976) – Durrell (1999):

As a boy, the author kept the **Scops owl** *(syczek)* at home on the island of Corfu in Greece. It is a small owl common in the Mediterranean, absent from Poland. When young Gerald obtained another species, the barn owl,

[71] But there is a generational difference. People born in the 1950s and earlier still remember a series of novels by Baroness Orczy (1865-1947) featuring the Scarlet Pimpernel and its numerous adaptations (e.g. a musical). The Scarlet Pimpernel is not likely to enjoy the same popularity among younger generations.

he had to make sure that it would not catch his Scops owl as prey, being three times larger (183). This information is preserved in the Polish version (183), where the diminutive *syczek* is replaced with the biggest of owls, ***puchacz*** ('eagle owl'). Certainly, eagle owls are not afraid of barn owls and rather than being substantially smaller they are twice as big as the latter[72]. In addition, the Scops owls are said to call "toink, toink" (41), the onomatopoeia thoughtlessly repeated in the Polish text (67), where a deep low-pitched hooting is expected. As in many other cases of 'domestication' of species names, translators must then diligently check the entire text for incongruities. Anecdotal evidence shows that they often lose control over the text and it becomes ridiculously incoherent and illogical.

IV. Fowles (1977: 178) – Fowles (1982: 214):

Conchis: 'For instance, since the age of eighteen I have refused to eat wild birds in any form at table, I would as soon eat human flesh as I would **an ortolan**, or a wild duck. ... And yet there I was, in black and white, refusing his *pate d'alouettes* and his truffled **woodcock**'.
In the Polish version an ortolan becomes ***przepiórka***, though woodcock is ***słonka***, as expected. It is true that in Poland there is no tradition of eating sparrow-sized wild birds. While a quail (*przepiórka*) on the table is not very surprising (although now illegal), an ortolan is. Perhaps that is the reason why the translator decided to replace 'ortolan', even if in Polish the bird has exactly the same name. Whatever the motivation, it strips readers of an opportunity to broaden their horizons and again impoverishes the diversity of nature Fowles keenly serves them at every step.

V. Fowles (1978: 344) – Fowles (2000: 413):

'Towhee', which is ***pipil***, becomes ***zięba*** in translation. The translator prefers to domesticate in this awkward way regardless of the obviously exotic context.

VI. Iwaszkiewicz (1969: 304) – Iwaszkiewicz (2002: 229):

'Bytuję jak **komar**, brzęczę nad trzęsawiskiem' – 'I only just exist like a **mayfly** hovering above the swamp'

---

[72] The above fragment comes from Kasprzak (2008: 182).

Although *komar* is 'gnat' or 'mosquito', here 'mayfly' functions well as a form of life to which one can compare one's fragile existence. The translator is consistent enough to replace *brzęczę* – 'I buzz' with 'hover' after substituting the buzzing gnat with the non-buzzing mayfly. The habitat remains the same as both families of insects occur in wet areas. But the question arises: Why are all these changes introduced? The text is not a piece of poetry, where rhyme, rhythm and other formal requirements may prevail over loyalty to the original imagery. It is difficult to find any motivation for the translator's manipulation other than a not-very-noble desire to have an outlet for her own self-expression.

Interestingly, if an English original of some text included 'gnat' and this would be rendered as ***jętka*** (mayfly), the effect would be most undesirable. While the name 'mayfly' sounds familiar to native speakers' ears, even if they know nothing about the creatures thus called, *jętka* belongs to scientific vocabulary and replacing the well-recognised *komar* with something alien to most Poles would make no sense.

VII. *The Monarch of the Glen*. 2000-2005. TV. BBC1. Ecosse Films:

In the subtitles to the film, '**grouse**' became ***kuropatwa*** ('partridge'), although the conversation concerned the red grouse, one of the icons of British culture (see chapter 3.2.). On assumption that it was conscious manipulation rather than sloppiness, we can interpret it as domestication, an attempt to give Polish viewers a familiar-sounding name of a gamebird. Undoubtedly, *pardwa*, which does not occur in Poland, does not meet the criterion of familiarity. However, the film did not present Polish reality and just as no one came up with the idea to make its characters pay in zlotys and eat *barszcz czerwony*, no one should have falsified other aspects of reality.

4.2.4.5. Corrections

When they occur, they are usually welcome, as in the following cases:

I. Lewandowska and Lewandowski (eds) (1994: 24):

Kaczka łyska – a **coot.** The coot, though a water bird, is not one of the duck species, as the Polish version suggests. It belongs to the family of rails (*Rallidae*), in Polish *chruściele*. Fortunately, this error is not repeated in translation.

II. Durrell (1973: 91) – Durrell (2001: 66):

'But as they [hairy frogs] were extremely fat, this long fast did not unduly worry me, for most **reptiles** can go for long periods without food and be none the worse for it.' –
'... a ponadto większość **gadów i płazów** może bez szkody dla siebie obejść się przez dłuższy czas bez jedzenia.'

Even Durrell himself managed to make a silly mistake. In this case, however, the translator has noticed it and 'reptiles' became 'reptiles and amphibians' to embrace the frogs in question, which are amphibians, and yet preserve the message that reptiles, like amphibians, can do without food for a long time.

III. Durrell (1976: 84) – Durrell (1999: 110)

'and the **carbon monoxide** given out by the spider would soak through the silky walls of its house' – 'a **dwutlenek węgla** wydalany przez pająka [topika] przesączał się przez jedwabiste ściany jego domu'
The watchful translator changed the impossible carbon monoxide *(tlenek węgla)* into carbon dioxide *(dwutlenek węgla).*

The problem is that the author's mistakes concerning nature terminology are rarely noticed by translators. If they are, the correction does not always meet the best standards, as the use of hypernyms for incorrect Polish names of six-spot-burnet or *Graphasoma italica* illustrates (cf. 4.2.4.1.2.). Unauthorised 'corrections' can be found in this work among the items listed under such categories as Additions, Omissions and Adopting the wrong translation strategy.

4.2.4.6. Changing the text's intertextuality

This error can be exemplified by Durrell narratives and *The Białowieża Forest saga* by Simona Kossak. In translation, against the will and effort of the author, the former become simple adventure stories. Innumerable errors in nature terminology, umpteen instances of the unauthorised domestication of names of species leading to a build-up of nonsense devalue the books as a source of information on the natural world. Disarming humour and hilarious stories may even persuade the Polish reader to support the cause of saving endangered species, of which Durrell was the champion, but in contrast to the readers of the original s/he does not obtain reliable knowledge on those species, exposed to recurrent inconsistencies and suspiciously incompatible names.

The translation of Kossak's book also contains numerous errors but they do not suffuse the text with ludicrous illogicalities as the translations of Durrell do. However, the translator strips the text of its cultural flavour. Kossak interlaces her standard Polish narration with Ruthenian-based regionalisms, Russian phrases and archaisms used to describe ancient professions and practices connected with the Białowieża Forest. She does it to give the reader a taste of the place – a mixture of Polish, Belorussian and Russian culture and to present its turbulent past stigmatised with years of Russian rule. The English text standardises the language and gets rid of culture-specific details and etymological explanations of Polish and local words. The author intersperses the narration with her comments on the human attitude towards the Białowieża Forest, animals and nature in general, some of them very critical of the Polish authorities. Not all of these found their way into the English version, mitigating the manifesto-like style of certain fragments. In conclusion, the book, though still quite subjective in style, appears to be more of a dispassionate and flavourless account of historical events and natural history of the Białowieża Forest than a saga.

Two subcategories of Hejwowski's meta-translation errors are excised due to the lack of illustrative material. Leaving two versions, characteristic rather of classroom translations, is an error eliminated by publishers, who do not want to pay for more than is necessary. Footnotes hardly ever occur in the sources studied in the present work but their extreme scarcity does not seem to deserve criticism. The bulk of data collected for this book, which concerns wrong and confusing names of species is not cited in the above taxonomy because it consists of entries which belong to the already most sizeable class of errors of mistaken interpretation. At the same time, it is also true that they escape unequivocal classification and all of them would have to be indexed as potential realisation errors. Doomed to inconclusiveness, such an exercise seems pointless.

4.2.5. Recapitulation

To close this part of the chapter, two noteworthy remarks can be made. Firstly, out of all categories, regadless of their hierarchical status, the most well represented are 'Misreading syntagms and wrong interpretation of verb frames', 'Misinterpreting scenes and scripts', 'Using hypernyms' (instead of lower-rank categories), 'Calques' and 'Target language errors'. Secondly, we have discovered that within two of the subtypes, a separate homogeneous group emerges. The random employment of most common dictionary equivalents (4.2.1.1.) can be subdivided into the uncritical use of most common, incorrect dictionary equivalents. Among misread syntagms and misinterpreted verb frames (4.2.2.1.) numerous instances of using *pars pro toto* appear. Typically, instead of the name of family or other higher taxa, the name of a specific species is given. As predicted, in many cases, assigning an item to just one particular category has proved impossible, which is reflected in brief comments pertaining to relevant entries. We can conclude that Hejwowski's classification helps enhance issues to which the translator who encounters nature terminology in any kind of text must develop sublime sensitivity. It also demonstrates the devastating effects of all sorts of mistranslation on the linguistic view of the world. Target language images differ from the source language pictures not in fine grain, the necessity of which nobody questions, but in essential – coarse grain components. Since the message recoverable from the overwhelming majority of categories eventually boils down to a warning against imagery distortion, the remainder of this chapter analyses it in more detail and with emphasis.

4.3. Imagery – errors

The following analysis of errors in terms of imagery distortions is meant as an alternative to classifications such as Hejwowski's. Its *raison d'être* is that it does not multiply categories and subcategories, which, when it comes to the crunch, turn out fuzzy, anyway. As seen above, they criss-cross each other's boundaries unpredictably, even at higher levels. It is difficult to determine when an error occurs due to misinterpretation and when because of faulty realisation. Neither is it easy to discern whether a wrong translation strategy has been adopted, or 'domestication' stems from a preposterous failure to find equivalents of species names. On the ensuing pages I re-examine six fragments of texts comprising errors placed in four basic categories of Hejwowski's taxonomy. The largest type of misinterpretation errors is represented by two examples – one taken from 'misreading syntagms and wrong interpretation of verb frames', the other from 'misinterpreting scenes and scripts'. Meta-translation errors are also illustrated with two case studies in acknowledgement of the diversity of their subcategories. The analysis of both single categories such as *dziuplaki* and more complex excerpts involves establishing a prototype and the degree of departure from it in mistranslation as well as considering in what way the properties of categories have been tampered with. Additionally, in order to see the distortions of the image of the world projected in translation from more than one angle I reach for Langacker's dimensions of imagery.

4.3.1. Errors of syntagmatic translation

Chylarecki and Sawicki (2003: 58 and 59) treat the foreign reader to the revelation that the common sandpiper is sometimes called 'squeaker'. The prototype of the category 'common sandpiper' is a medium to small dark-backed wader uttering characteristic high-pitched shrill three-note alarm calls as it flies off. British bird watchers obviously coin abbreviated names or nicknames for birds[73] but they cannot possibly be calques of corresponding alternative names circulating among Polish birders. The common sandpiper can be a 'bobbob' or 'teetering' – both suggestive of its habitual bobbing of head and tail, but not a 'squeaker', which calques the Polish *piskliwiec* (*pisk* – a squeak, *piszczeć* – to squeak). Apparently, its voice is not perceived by most Brits as squeaking. If at all, it could occupy a position in the extreme periphery of the category. The prototypical squeaks are produced by mice, rats or guinea pigs. In English the category of squeakers is capacious enough without this new contribution: 'a noisemaker or a toy instrument that sqeaks', 'young racing pigeon', 'young pig', 'a rat cangaroo of the genus Bettongia', 'an African freshwater catfish' or '*Strepera versicolor* – an Australian bird', 'a contest won by a small margin', 'an informer or spy working for the police'. The translator extended the category of entities that squeak, dwelling heavily on the Polish verbalisation of non-linguistic sounds, which does not distinguish between loud and shrill noises and relatively quiet, high-pitched voices. They are all termed *pisk* qualified with the adjectives 'loud', 'quiet', 'shrill', etc., expressed through derivatives (*popiskiwanie* – 'quiet recurrent *pisk*') or via less conventionalised, onomatopoeic notations. This is what is done in reference to the common sandpiper in English – it does 'tee-wee-wee' or utters three-note shrill calls, maybe even emits three-note shrieks but by linguistic convention it does not squeak, tweet, twitter or scream (although at a very general level this last lexical item occurs in folk usage). In terms of dimensions of imagery we can say that by writing 'squeaker', the translator profiled the wrong element from the vantage point that the prospective receptors do not share and therefore can hardly appreciate. The auditory image evoked in their minds departs greatly from the prototype. If they do not have any prototypical common sandpiper voice to rely on, on the basis of 'squeaker' they can form one. This is one of the ways in which people develop false conceptualisations of nature. On assumption that foreign readers expect to learn some cultural peculiarity from a publication in question 'Piskliwiec –

---

[73] See e.g. http://www.birdforum.net/archive/index.php/t-14263.html%20 - 87k accessed 4 February 2009.

bo tak go się skrótowo nazywa' could be replaced with a footnote or bracketed information: 'Polish birders call it *piskliwiec* for short'. People willing to know the official Polish name can find it in any guide: *brodziec piskliwy*, the jargon name available from far fewer sources. In this way, we do not overstretch an existing category but invite the receptors to familiarise themselves with a totally new one. The linguistic view of the world rather than being distorted stands a chance of becoming broader in perspective.

4.3.2. Misinterpretation errors

4.3.2.1. Misread syntagms

To begin with, we need to compare images emerging from an extract of the original text by Czesław Miłosz and its translation.

> Być może kiedyś było tutaj jezioro, ale na jego dawnym dnie teraz rozpościerały się łąki z ostrą wikswą i dalej, przed nimi, **mszaryna** – kępy, na których rosły karłowate sosenki, tu i ówdzie chaszcze splątanej **łozy**. ... Jest to królestwo zapachu. Z mchu sterczą tu krzaki bahunu – *Ledum palustre* – z ich wąskimi skórzanymi liśćmi, niebieskie jagody pijanic, duże jak gołębie jajka, dojrzewają w cieple powietrza przesyconego parą. ... Młode cietrzewie znajdują tu dość jedzenia, pod kierownictwem matki ...' (Miłosz 1989: 214).

The text contains several nature terms: two official, four non-standard, and a number of standard-inexpert ones, some of which call for explication. *Wikswa*, the word unknown to contemporary Poles, comes from Lithuanian *viksva* – 'sedge' – *turzyca*. *Mszaryna* has its standard form *mszar* – raised bog or bog woodland. The word enhances one important component of the profile of well-preserved raised bogs and bog woodlands – cushions of sphagnum mosses (*mszysty* – 'of moss'). *Karłowate sosenki* means dwarf/stunted pines, *chaszcze* – a thicket (of shrubs), *łoza* – grey willow. *Bahun*, a regional name, is officially and commonly known as *bagno zwyczajne*, *pijanica* ('drunkard' – as it is believed to have slightly narcotic properties)[74] has a more widespread unofficial usage but for specialists it bears the name of *borówka bagienna* – bog bilberry. *Cietrzew* stands for the black grouse and simultaneously for the male of the species. The scene painted by the author shows sedge meadows developed in the

[74] This is a folk truth, which has been scientifically disproved (Grochowski and Grochowski 1994: 162).

former lake basin, bordering on a raised bog in the process of transformation into bog woodland. Here and there (most probably in the meadows rather than bogs) grey willow thickets appear in clumps. In the bogs grow the marsh tea and bog bilberries, which make excellent feeding grounds for the black grouse chicks led by their mother. The published translation is quoted below for comparison:

> Once a lake, its former basin was now a sprawling meadow covered with sharp sedge and, up ahead, with moss, **thick with pine scrub** and clumps of tangled withes. ... The **moorland** was a kingdom of fragrance. The moss sprouted bushes of marsh tea (*Ledum palustre*) with narrow leathery leaves, and blue bog bilberries the size of pigeon eggs thrived on the warm, vaporous air. ... The **moor** made a good feeding ground for the young **blackcocks**, still under their mother's guardianship ... (Miłosz 1981: 235)

First and foremost, the meadows do not border on a bog but on moorland – an upland habitat overgrown with heather, grass and gorse. Moorlands can be boggy, so the marsh tea and bog bilberries in it would not surprise if it were not for the omnipresent moss thick with pine scrub. A moor covered with thick pine scrub is no longer a moor. Its typical vegetation demands more sunlight. If it is moorland, why does the heather not find its way into the description? Where are the hills? The gorse can be skipped – after all it does not occur in the east of Europe. While Miłosz dots the landscape with clumps of grey willow, the translator tells us about unspecific tangled withes (flexible slender twigs) – *witki* in Polish. Finally, he makes his narrator even more omniscient than the author, informing us that all the chicks under the guardianship of the black grouse hen are cocks. One could also speculate about substituting 'marsh tea' and 'bog bilberry' with less standard equivalents but that might raise controversy. *Ledum palustre* has numerous alternative names in English and it is difficult to establish which sound more obscure to most native speakers. It can be inferred from an Internet search and from my first hand experience that at least three have a roughly identical statistical distribution. The others listed in Anioł-Kwiatkowska (2003) seem hardly ever used and look scientific rather than folk. The bog bilberry apparently enjoys a better reputation in Anglo-Saxon culture and no equally pejorative label as *pijanica* can be traced. Regional names abound but since they all follow the same pattern as the official one – bog + X-berry, we cannot take responsibility for recommending them in this particular situation.

The Polish prototype of the scene has very little to do with the one that can be envisaged on the basis of the translation. Some prototypical properties are almost opposites – bog woodland in the Lithuanian climate grows in lowland areas, the UK's moors in uplands. Drier bogs are naturally encroached by pines – not by scrub, though. British/Irish moors are never covered with expansive stunted pines because of a huge red deer population and their purposeful removal to facilitate the growth of heather. In contrast to bogs, moors may or may not be boggy, some can be relatively dry. The boundaries of the category presented to the foreign receptors follow significantly different lines. As discussed in earlier chapters, frontiers of categories such as bog, bog woodland and moor are fluid despite efforts to specify them better. Here, they are made far more fuzzy than sanctioned by both folk and specialist usages. Some core exemplars of 'bog woodland' are shifted to 'moor', where they do not normally enter the edges of the periphery, while basic elements of 'moor' do not even linger in the background of the categorial set. Undoubtedly, the stretchability of categories is unforgivably abused. The elucidation of another problem requires turning to the parameters of imagery. At the level of schematicity and, inexorably, of perspective the translator tilts the overall tone towards ornithology rather than botany. He overspecifies, detailing the sex of young black grouse and underspecifies, mentioning clumps of tangled withes. He foregrounds (inadvertently) the characteristic features of moor such as [HEATHER COVER], [TREELESSNESS], without the relegation of defining properties of bog woodlands to the background. To the bewilderment of informed readers [PROMINENCE OF TREES], [MOSS COVER] and [WATERLOGGING] stay put. The mess is caused by a clumsy manipulation with boundaries and membership of several linguistic categories. The linguistic picture of the world offered to the foreign recipient blurs the extralinguistic reality.

4.3.2.2. Misinterpreting scenes and scripts

Here is a scene painted by Gerald Durrell:

> …a black-and-white wattled plover that **flapped away upstream**, tittering imbecilically to itself, with **its feet grazing the water**, and long yellow wattles flapping absurdly on each side of its beak. (Durrell 1975a: 28)

And here is a scene presented by the translator:

> czarno-białe siewki, **człapiące w górę strumienia** i chichoczące idiotycznie do siebie, **nogami grzebały w wodzie**, a ich długie, żółte korale trzepotały absurdalnie po obu stronach dzioba.' (Durrell 1991: 37)

Anyone well-acquainted with both languages can see that these two project quite dissimilar images. In the Polish version the subject is not a single wattled plover but a flock of unspecific plovers. They did not flap away but plodded (sic!) upstream. Their feet were not grazing the water as they flew but they 'raked' the water and, by implication, the muddy bottom as if probing for food. The English version vibrates with dynamism, while the Polish one is nearly static. Both texts share two messages: the bird titters imbecilically and has yellow wattles. Leaving aside the question of why the translator multiplied the plovers, let us concentrate on the prototypes of the main character. In Polish we have approximately sixty species to choose from, although, admittedly, the yellow wattles indicate the specific *siewka – czajka płowa*, certainly not the first species that comes to the Polish mind on hearing the term *siewka*. In English we do not have to guess, knowing outright the name of this exotic African species – the category's boundaries are considerably less fuzzy. The creature in the translation is unlike any plover – awkward and heavy, behaving like a cross between a domestic chicken (scratches the ground with its feet) and a swan (walks clumsily, sort of plodding). The English text gives us a perfectly credible account of a plover's appearance and behaviour – it is graceful, light and swift. The two versions point to different profiles, the Polish version falsifying the reality. The linguistic view of the world recklessly distorts a particular fragment of reality rather than attempting to reflect it faithfully (however imperfect it is bound to be) in the lexicon. Last but not least, as the level of schematicity increases, the Polish version changes the perspective, too. The observer becomes less of an expert.

4.3.3. Realisation errors

As in many other cases, the mistranslation discussed beneath may not belong to the class of realisation errors at all but stem from the wrong interpretation. I have decided to treat it as failure to find the suitable English equivalent, caused probably by the translator's complacency about the correctness of one of two possibilities that came to his head. This inhibited all the control mechanisms, which would otherwise have made him realise that the hastily rejected possibility was actually better in spite of a lack of exact correspondence of meaning with the Polish term.

Sawicki (2003b: 64) writes: 'Są one niezbędne dla niektórych dziuplaków, które nie potrafią wykuć dziupli w żywym, twardym drewnie, jak np. sikory: czarnogłowa i uboga.' In translation we receive: 'They [dead trees] are very much needed by certain hollow drilling birds such as the willow tit and marsh tit, which are unable to drill [a] hole in the living hard wood.' (Sawicki 2003b: 65).

To recollect the definition formulated on p. 240 above, *dziuplaki* are hole nesters – birds which nest in holes, not only in trees. Some of them excavate their own holes in hard living wood, some in dead wood but some simply occupy natural holes (in trees, rocks, buildings and earth) or those made by other species. Both the Polish and English lexical items agree as to the maximal scope of the category and the prototype. The maximal scope encompasses species of various sizes and habits (from as large as goosanders, goldeneyes and owls through woodpeckers, rollers, hoopoes, starlings to nuthatches, tits, and flycatchers), the prototype is in all likelihood one of the woodpeckers. Hollow drilling birds certainly are *dziuplaki* but the prototype becomes here the only, though collective, member. To be precise, it should be sought only among a tally of woodpecker species, a fairly big category but radically smaller than 'hole nesters'. For most Poles and Brits it could be the great spotted woodpecker (*dzięcioł pstry duży*), most numerous and tolerant of human presence. By narrowing down the scope the translator made the category less fuzzy, but doing so, fell into the trap of overspecification and wrong profiling. Thus incurred illogicality contributed to the distortion of the linguistic view of the world in the English version. Perplexed readers wonder how the creatures that can drill cannot do it in any type of wood. The category 'hole nesters' allows for the membership of 'non-drillers', which remove soft dead matter scooping out a hole. The beaks of tits are too weak to drill but they actively make their holes, which must have misled the translator during his search for the right term.

4.3.4. Meta-translation errors

4.3.4.1. Inappropriate choice of translation technique

The translator of Sawicki (2003b) renders *Lasy Państwowe* (68) as National Forests (69), ignoring the existence of the official translation of the name of this powerful institution: the State Forests, with the State Forestry Authority at the head. In this way he violates an important rule: always check names of official organisations, institutions, other public bodies and administrative units before coining your own. Concurrently, he minimises the categorial scope and changes its prototype. The Polish 'State Forests' category has the following properties: 1. covers the majority of all woodland in the country, access to them being free except in special situations; the rest is private woodland 2. embraces forests of great ecological importance as well as plantations, all (to varying degrees) managed for timber, recreation and wildlife 3. areas in their care are forests, largely of long standing, only occasionally interspersed with grassland, mires or other open habitats; towns and villages do not form a part of the institution's land. The corresponding institution which manages the state forests in the UK is the Forestry Authority and if we treat it as a category, it has remarkably different features: 1. covers the relatively small portion of woodland in the country which is not in private hands. 2. most of it is for commercial purposes, although management for amenity and wildlife has been increasingly adopted. 3. areas in their care are small and scattered, there are little old-growth stands.

In Britain the term 'National Forest' brings totally different associations. Since the 1990s a woods planting project has been implemented on both public and private spaces in the Midlands with a view to collectively forming the National Forest. It has aimed to increase the forest cover in the UK because it is far below the European average and sustainability thresholds. The National Forest encompasses all landscape within its confines, together with towns and villages and has a complex ownership and management structure, only coordinated at a national level. On assumption that the above facts are capable of shaping the overall image of the linguistic category of national forests everywhere abroad in the mind of a typical British person, we can say that she or he would envisage Polish State Forests as a category substantially poorer in membership with differently drawn frontiers. If we look at it more optimistically, that the prototype may arise from a less parochial perspective, 'National Forests' sounds like national parks – gems of a given country's nature, managed pimarily for the conservation of natural features and for public enjoyment. Again, a huge proportion of category members would have to move out. In both interpretations the translator is guilty of wrong profiling, overlooking existing conventionalisations and failing to realise the altered perspective of the foreign receptor. It goes without saying that errors of lexical categorisation impose a distorted view of the real world. Although the British may be perfectly aware that state forests in other countries look different in many respects and may allow for a variety of prototypes and members to build their image, their imagination should not be led astray by haphazard choices of the translator.

4.3.4.2. Using hypernyms

Fowles populates his novels with a plethora of species names. Some of them attain a symbolic status, others at every step underline the unavoidability of interaction between humans and countless lifeforms. Therefore using hypernyms in translation, where the author enthusiastically indulges in fine grain depiction does injustice to him, the message of the original and the general climate of his books. Let us take a closer look at one of many such manipulations:

> It had brought out swarms of spring butterflies, those brimstones, orange-tips and green-veined whites … (Fowles 1996: 160)

> Przywitał roje motyli wiosennych, siarkowożółtych z pomarańczowym obrzeżeniem i zielonymi żyłkami. (Fowles 2003: 150)

Fowles wants the reader to enjoy a spring scene with three distinct species of butterflies in the foreground. Brimstones are so called because of their bright yellow colour (brimstone is the obsolete word for sulphur). Orange tips have white uppersides to their wings. The males have orange-tipped forewings and hence the name. The underside hindwings of otherwise white females of green-veined whites are pale yellow with the veins highlighted by black scales giving a greenish tint, which explains the name. In Polish, we get a very impressionistic vision of a swarm of butterflies, each of which is at the same time sulphur-yellow, orange-fringed and green-veined. Three separate entities become rolled into one fancifully coloured being. During this brutal 'merger' operation the absent-minded translator smeared orange along the edges of the entire wing surface instead of solely on their upper tips. The more respectful translator would render the cited fragment as follows:

> Przywitał roje motyli wiosennych, latolistków cytrynków, zorzynków rzeżuchowców i bielinków bytomkowców.

The Polish reader would find out a little less about the patterns and distribution of colour on the butterflies' wings, only that the first of the three is lemon yellow and the third white. The name of the second refers to *zorza* – the phenomenon of the sky's gradual changing of colour immediately before, during and soon after sunrise and sunset. Considering the broad spectrum of hues appearing at such times, the clue is not very helpful. But the same reader would probably marvel at the richness of nature, seeing three different species named in such intriguing ways. And that is a far more desirable effect on the part of the author than the appreciation of aesthetic appeal of a colourful nameless swarm. In this more faithful translation the perspective does not change – it is still the one of an expert, not the vantage point of a colour-sensitive dilettante. As in the English text, there are three categories, each possessing a separate prototype, not one and vague at that. The degree of schematicity decreases in comparison with the published translation but agrees in the level of specification with the original. Three distinct figures emerge from the ground rather than one melting into it like snow in a pool of water at high temperature. The view of the world projected through language also remains identical, certainly to the point categorisation as a cognitive phenomenon can be reconstructible in different languages and minds. For the very reason that such reconstructions are in themselves difficult the translator must not re-draw the borderlines around categories and throw their members in or out simply as fancy dictates.

4.4. Recapitulation

As can be seen, an imagery-based analysis is applicable across Hejwowski's taxonomy of errors. It seems that viewing nature terminology and the contexts it is embedded in as verbalised categories, whose members are manipulated with in the translation process, lends itself to a very sensitive study. These categories have been examined in respect of prototypicality, dimensions of imagery and the projected linguistic view of the world confronted with the extralinguistic reality. The departures from the prototype, unreflective or poorly motivated boundary shifts, unjustified changes of the degree of conventionalisation, perspective, salience and schematicity, when brought to the translators' attention, could have a sobering effect. A sober look helps see further afield but it can also act as a depressant, which fosters untranslatability scaremongering. To avert the danger, this chapter has gone beyond an eye-opener mission, offering practical solutions.

CONCLUSION

The above study of categorisation corroborates the conclusion drawn by Edward T. Hall: '… in his strivings for order, Western man has created chaos by denying that part of his self that integrates while enshrining the parts that fragment experience' (cited in Rheingold 1988: 137). As living in chaos is maddeningly uncomfortable, we are doomed to perpetually seek order if not by fragmentation then by integration of experience. If concepts and terms undergo continuous re-defining and re-naming, little seems entrenched enough to be considered definitely correct or faulty. What emboldens us nevertheless to pass judgements of correctness when it comes to the differentiation and then translation of verbalised categories stems, *inter alia*, from the following conclusion:

> Can knowledge ever be secure? Of course it can. … Our knowledge of the existence of cells seems secure, as secure as any knowledge is likely to be. Nonetheless it is human knowledge based on human understanding, not on any neutral, or God's-eye-view, understanding. There is no such thing as a neutral way to understand things. But as long as our human understanding remains stable, it is possible for our knowledge to be secure. (Lakoff 1987: 299-300)

Languages cut up reality in various ways, experts distinguish categories on the basis of often mutually exclusive criteria, but in a given context the number of possible divisions, or interpretations of a given fragment of reality can be, and usually has to be curtailed. Otherwise, communication would not take place. The recognition of contextual clues acts as a limiting factor and consequently as a key to successful communication. Due to perceptual differences among conceptualisers, such a recognition may be deficient. That legitimises definite correctness judgements and the reduction of the quantity of valid options. In other words, it enables us to achieve a degree of objectivity and confidence in criticism. Underpinned by thus derived legitimacy, this work has investigated the limits to the fuzziness of category boundaries, to the relativity of prototypes and to the liberty of manipulation with the parameters of imagery such as salience, perspective and schematicity. As, hopefully, sufficiently demonstrated in Chapters Three and Four, such limits can be found, although they are always a matter of degree.

By choosing cognitive analysis techniques, I admit that more reductive and rigid methods would counter-productively lead to the deepening of chaos. Cognitivism allows us to both fragment and integrate our mental and external reality. It tightly binds together linguistic and extralinguistic worldviews. The discussion of the phenomena of nature and their various encodings in two languages – Polish and English – cannot help but be anchored in both realities. I have adhered to the stance that the equivalence of two texts is the equivalence of their mental representations and that it hinges on the successful mapping of one language onto the other in all pertinent aspects of Langacker's imagery. Although the translator has at his/her disposal only linguistic tools, they are but a cog in a sophisticated mental machinery responsible for the success of the mapping. S/he has no control over them and must humbly focus on the matching and adjustment of verbalised categories but with a view towards their chances of molding desirable mental images. It flows from this that the search for suitable lexical and grammatical items must be paralleled by the awareness of natural and cultural conditions, which influence the reception of the world encapsulated in the imperfect symbolic structures of the target language. Without this pragmatic knowledge, verbalisations, though syntagmatically correct, may interpretatively turn out to be a total flop.

Although I have borrowed and adapted certain theoretical constructs, theorising has not been put at the centre of attention. I have aimed at boosting the translators' intuition and wanted to steer clear of having to admit eventually, like Tabakowska in one of her articles, that 'all this complicated analysis has brought us to the point to which a sensitive reader and an expert translator are inevitably led simply by intuition' (1998: 182-183, translation mine).

This book emphasises the role of extralinguistic knowledge of the natural world. In order to be translatable intralingually, in ways such as those elaborated on in Chapter Two, it needs to be internalised, tamed and eventually it must become intimate. Only then can one translate nature into a foreign language. The translator who feels the same as Eva Hoffman in the quotation below cannot do a good job: 'How could I say "darling", or "sweetheart", when the words had no fleshy fullness, when they were as dry as sticks?' (Hoffman 1989: 245).

Hoffman felt that about the English language. Before she absorbed enough of it, the language of endearment was Polish for her. But a native Pole faced with the task of translating a text teeming with nature terminology can ask exactly the same question: how can I translate into English *łąka trzęślicowa*, *mszar*, or the description of the capercaillie's song if they have no fleshy fullness to me and if I cannot trust dictionaries? The answer is fortunately more optimistic than in the case of trying to find one's feet in a new culture: consult somebody for whom these phenomena and their verbal representations are not as dry as sticks. The alternative is: ask, read, observe, get involved in nature conservation and research and never tire of it. But that takes as much time and heart as merging into an alien culture.

Thus, while a linguist can contribute to the improvement of the quality of translation through his sensitive analytical tools, it takes a naturalist to ensure that a copy produced as an outcome of translation is maximally faithful. The label 'naturalist' does not imply a professional. A hobbyist, in whom the folk tendencies to seek similarity and an expert's drive to dissect strike a balance, and who is also a translator, may prove the most sensible copyist. 'Sensible' means here that the copyist is neither too meticulous nor too blind to detail but that he/she responds to the demands imposed by a specific context. Both extremes, when in the wrong context, promote chaos rather than integration and better communication.

REFERENCES

Translation studies and linguistics

Anusiewicz, Janusz. 2004. Problematyka językowego obrazu świata w poglądach niektórych językoznawców i filozofów niemieckich XX wieku. In *Językowy obraz świata*, edited by J. Bartmiński. Lublin: Wydawnictwo UMCS, pp. 261-289.

Bartmiński, Jerzy. 2004. Punkt widzenia, perspektywa, językowy obraz świata. In *Językowy obraz świata*, edited by J. Bartmiński. Lublin: Wydawnictwo UMCS, pp. 103-120.

Bartmiński, Jerzy, and Ryszard Tokarski, eds. 1998. *Profilowanie w języku i w tekście*. Lublin: Wydawnictwo UMCS.

Dougherty, Janet, W. D. 1999. Wyrazistość i względność klasyfikacji. Translated by Wojciech Kubiński. In *Amerykańska antropologia kognitywna*, edited by M. Buchowski. Warszawa: IK, pp. 160-185.

Felber, Helmut, and Gerhard Budin. 1994. *Teoria i praktyka terminologii*. Translated by Czesława Schatte. Warszawa: Wydawnictwa Uniwersytetu Warszawskiego.

Grzegorczykowa, Renata. 2004. Pojęcie językowego obrazu świata. In *Językowy obraz świata*, edited by J. Bartmiński. Lublin: Wydawnictwo UMCS, pp. 39-46.

—. 1996. Filozoficzne aspekty kategoryzacji. In *Językowa kategoryzacja świata*, edited by R. Grzegorczykowa and A. Pajdzińska. Lublin: Wydawnictwo UMCS, pp. 11-26.

Gutt, Ernst-August. 2004. Interpretive core activities in translation: Inferring, designing and comparing interpretations. In *Relevance studies in Poland. Volume I* , edited by Ewa Mioduszewska. Warsaw: Institute of English Studies, University of Warsaw, pp. 157-178.

—. [1991] 2000. *Translation and relevance*. Manchester & Boston: St. Jerome Publishing.

Habrajska, Grażyna. 1996. Kategoryzacja i klasyfikacja – potoczne i naukowe widzenie świata (analiza badań ankietowych dotyczących kategoryzacji nazw roślin). In *Językowa kategoryzacja świata*, edited by R. Grzegorczykowa and A. Pajdzińska. Lublin: Wydawnictwo UMCS, pp. 221-241.

Hejwowski, Krzysztof. 2004a. *Translation: A cognitive-communicative approach*. Olecko: Wydawnictwo Wszechnicy Mazurskiej.

—. 2004b. *Komunikacyjno-kognitywna teoria przekładu*. Warszawa: Wydawnictwo Naukowe PWN.

Hoffman, Eva. 1998. *Lost in translation*. London: Vintage.

Hunn, Eugene. 1999. Czynnik utylitarny w ludowych klasyfikacjach biologicznych. Translated by Joanna Dankowska. In *Amerykańska antropologia kognitywna*, edited by M. Buchowski. Warszawa: IK, pp. 73-97.

Kardela, Henryk. 2004. Ogdena i Richardsa trójkąt uzupełniony, czyli co bada gramatyka kognitywna. In *Językowy obraz świata*, edited by J. Bartmiński. Lublin: Wydawnictwo UMCS, pp. 15-37.

Kasprzak, Wojciech. 2008. 'Prime eagle' and 'Poor tit' in a quagmire of translation – on processing cost savers. In *Relevant Worlds: Current perspectives on language, translation and Relevance Theory*, edited by E. Wałaszewska, M. Kisielewska-Krysiuk, A. Korzeniowska and M. Grzegorzewska. Newcastle upon Tyne: Cambridge Scholars Publishing, pp. 171-187.

Korzeniowska, Aniela. 1998. *Explorations in Polish-English mistranslation problems*. Warszawa: Wydawnictwa Uniwersytetu Warszawskiego.

Lakoff, George. 1987. *Women, fire, and dangerous things. What categories reveal about the mind*. Chicago-London: The University of Chicago Press.

Langacker, Ronald W. 2009. *Investigations in cognitive grammar*. Berlin. New York: Walter de Gruyter.

—. 1991. *Concept, image, and symbol. The cognitive basis of grammar*. Berlin: Mouton de Gruyter.

—. 1987. *Foundations of cognitive grammar. Theoretical prerequisites*. Vol. I. Stanford: Stanford University Press.

Lewandowska-Tomaszczyk, Barbara. 1996. Uniwersalizm a relatywizm na tle nowych teorii kognitywnych. In *Językowa kategoryzacja świata*, edited by R. Grzegorczykowa and A. Pajdzińska. Lublin: Wydawnictwo UMCS, pp. 47-69.

Maćkiewicz, Jolanta. 2004a. Kategoryzacja a językowy obraz świata. In *Językowy obraz świata*, edited by J. Bartmiński. Lublin: Wydawnictwo UMCS, pp. 47-55.

—. 2004b. Wyspa – językowy obraz wycinka rzeczywistośći. In *Językowy obraz świata*, edited by J. Bartmiński. Lublin: Wydawnictwo UMCS, pp. 193-206.

—. 1996. Od chaosu do porządku, czyli o pewnym typie kategorii naukowych. In *Językowa kategoryzacja świata*, edited by R.

Grzegorczykowa and A. Pajdzińska. Lublin: Wydawnictwo UMCS, pp. 243-253.

—. 1992. Leksyka ogólna a terminologia – dwa sposoby opisywania świata (na przykładzie polskich i angielskich nazw granicy między wodą a ziemią). In *Prace filologiczne* tom XXXVII. Warszawa, pp. 221-228.

—. 1991. *Nienaukowy i naukowy obraz morza*. Gdańsk: Wydawnictwo Uniwersytetu Gdańskiego.

Marczewska, Marzena. 2002. *Drzewa w języku i w kulturze*. Kielce: Wydawnictwo Akademii Świętokrzyskiej.

Muszyński, Zbysław. 1996. Światy za słowami. Ich natura i porządek. In *Językowa kategoryzacja świata*, edited by R. Grzegorczykowa and A. Pajdzińska. Lublin: Wydawnictwo UMCS, pp. 27-45.

Newmark, Peter. 1988. *Approaches to translation*. London: Prentice Hall Int (UK) Ltd.

Pajewska, Ewa. 2003. *Słownictwo tematycznie związane z lasem w kontekście badań nad językami specjalistycznymi*. Szczecin: Wydawnictwo Naukowe Uniwersytetu Szczecińskiego.

Pieńkos, Jerzy. 1993. *Przekład i tłumacz we współczesnym świecie*. Warszawa: PWN.

—. 2003. *Podstawy przekładoznawstwa*. Kraków: Zakamycze.

Pilegaard, Morten. 1997. Translation of medical research articles. In *Text typology and translation*, edited by Ann Trosborg. Amsterdam/ Philadelphia: John Benjamins, pp. 159-184 .

Rheingold, Howard. 1988. *They have a word for it: a lighthearted lexicon of untranslatable words and phrases*. New York: Tarcher.

Robinson, Douglas. 1997. *Becoming a translator*. London: Routledge.

Sager, Juan. C. 1997. Terminology. Applications. Theory. In *Routledge encyclopaedia of translation studies*, edited by Mona Baker. London: Routledge, pp. 251-262.

Skowronek, Barbara. 2001. O nauczaniu języków specjalistycznych. In *Języki fachowe, problemy dydaktyki i translacji,* edited by A. Kątny. Olecko: Wszechnica Mazurska, pp. 115-121.

Strutyński, Janusz. 1972. *Polskie nazwy ptaków krajowych*. Wrocław: Zakład Narodowy im. Ossolińskich.

Szczekocka-Augustyn, Aleksandra, Beata Wereszczyńska, and Teresa Zagrodzka. 1996. O kategoriach rozmytych (na przykładzie rozumienia nazw fauny, flory i artefaktów). In *Językowa kategoryzacja świata*, edited by R. Grzegorczykowa and A. Pajdzińska. Lublin: Wydawnictwo UMCS, pp. 255-271.

Tabakowska, Elżbieta. 1999. *O przekładzie na przykładzie*. Kraków: Znak.

—. 1998. Profilowanie w języku i w tekście – perspektywa językoznawcy, tłumacza i poety. In *Profilowanie w języku i w tekście,* edited by J. Bartmiński and R. Tokarski, Lublin: Wydawnictwo UMCS, pp. 167-184.

—. 1995. Językoznawstwo kognitywne w teorii i praktyce przekładu: oleodruk i symfonia na dwa fortepiany. In *Między oryginałem a przekładem I*, edited by J. Konieczna-Twardzikowa and U. Kropiwiec. Kraków: Universitas, pp. 31-41.

—. 1993. *Cognitive linguistics and poetics of translation*. Tübingen: G. Narr.

Taylor, John. R. 1995. *Linguistic categorization. Prototypes in linguistic theory*. Oxford: Clarendon Press.

Tokarski, Ryszard. 1996. Ramy interpretacyjne a problemy kategoryzacji (przyczynek do tak zwanej definicji kognitywnej). In *Językowa kategoryzacja świata*, edited by R. Grzegorczykowa and A. Pajdzińska. Lublin: Wydawnictwo UMCS, pp. 97-113.

—. 1995. Tło kulturowe a znaczenia jednostek leksykalnych. In *Kreowanie świata w tekstach*, edited by A. M. Lewicki, and R. Tokarski. Lublin: Wydawnictwo UMCS, pp. 39-58.

—. 1993. Słownictwo jako interpretacja świata. In *Encyklopedia kultury polskiej XX wieku, vol. 2*, edited by J. Bartmiński. Wrocław: "Wiedza o Kulturze": Instytut Filologii Polskiej UMCS, 1993, pp. 335-362.

Wierzbicka, Anna. 1997. *Understanding cultures through their key words: English, Russian, Polish, German and Japanese*. Oxford: OUP.

Wiśniewska, Halina. 2001. *Język polski na ziemiach ruskiej i lubelskiej w Rzeczypospolitej szlacheckiej XVI-XVIII w.* Lublin: Wydawnictwo UMCS.

Zinken, Jörg. 2008. Linguistic picture of the world or language in the world: Metaphors and methods in ethnolinguistic research. *Etnolingwistyka* 20: 51-62.

Nature terminology: analytical studies, reference materials and dictionaries

Anioł-Kwiatkowska, Jadwiga. 2003. *Wielojęzyczny słownik florystyczny*. Wrocław: Wyd. Uniwersytetu Wrocławskiego.

Bick, Walter, Allan Robertson, Renate Schneider, Siegfried Schneider, and Piotr Ilnicki. 1976. *Słownik torfoznawczy niemiecko-polsko-angielsko-rosyjski.* Warszawa: Państwowe Wydawnictwo Rolnicze i Leśne.

Blamey, Marjorie, Richard Fitter, and Alastair Fitter. 2003. *Wild flowers of Britain and Ireland*. London: A & C Black.

Bragg, Olivia, and Richard Lindsay, eds. 2003. *Strategy and action plan for peatland conservation in Central Europe*. Wageningen: Wetlands International.

Brown, Alan, ed. 1992. *The UK environment*. London: HMSO.

Brückner, Aleksander. 1996. *Słownik etymologiczny języka polskiego*. Warszawa : Wiedza Powszechna.

Bunce, Michael. 1994. *The countryside ideal. Anglo-American images of landscape*. London and New York: Routledge.

Busse, Przemysław, ed. 1991. *Mały słownik zoologiczny. Ptaki*. Warszawa: Wiedza Powszechna.

Carton de Wiart, Adrian. 1950. *Happy odyssey*. London.

Couwenberg, John. 2005. *Mires of the world*. Power point presentation. University of Greifswald.

Cowie, A. P., ed. 1989. *Oxford advanced learner's dictionary*. Oxford: OUP [OALD].

Czekierda, Krzysztof. 1992. *Słownik ekologiczny polsko-angielski*. Warszawa: Terra Sana.

—. 1995. *Słownik ekologiczny angielsko-polski*. Białystok: Wyd. Ekonomia i Środowisko.

Decyzja nr 5 Dyrektora Generalnego Lasów Państwowych z dnia 30.01.2007 w sprawie metodyki inwentaryzacji siedlisk i roślin. Załączniki 1, 2, 3.

Dobrowolski, Kazimierz, ed. 1991. *Ptaki Europy*. Warszawa: Elipsa.

Drive Publications Limited, eds. 1981. *Book of the British countryside. Great Britain*: DPL.

Dunn, Euan (consultant). 1990. *The ever-changing woodlands. The living countryside*. London, New York, Montreal, Sydney, Cape Town: The Reader's Digest Association Ltd.

Dyduch-Falniowska Anna, Jacek Herbich, Maria Herbichowa, Wojciech Mróz, and Joanna Perzanowska. 2002. Krótka charakterystyka typów siedlisk przyrodniczych o znaczeniu europejskim, występujących w Polsce. Wdrażanie koncepcji Natura 2000 w Polsce w latach 2001-2003. Materiały dla Wojewódzkich Zespołów Realizacyjnych. Kraków – Gdańsk. Maszynopis.

Dyduch-Falniowska, Anna, and Katarzyna Zając, eds. 1996. *CORINE biotopes w integracji danych przyrodniczych*. Kraków: Wyd. Instytutu Ochrony Przyrody PAN.

Dziubiak, Marta. 2004. Collection of the Genus Malus Mill. in the Botanical Garden of the Polish Academy of Sciences in Warsaw. *Journal of Fruit and Ornamental Plant Research*, vol. 12: 121-128.

Gibbons, Robert (consultant). *The evergreen world. The living countryside*. 1991. London: Reader's Digest Association Ltd.

Gibbons, Bob. 1999. *Insects*. London: HarperCollins*Publishers*.

Gooders, John. 1998. *Birds of Britain and Europe*. London: HarperCollins*Publishers*.

Gove, Babcock Philip, ed. 1993. *Webster's third new international dictionary*. Cologne: Konemann.

Grochowski, Wiesław, and Andrzej Grochowski. 1994. *Leśne grzyby, owoce i zioła*. Warszawa: PWRiL.

Hall, J. E., and K. J. Kirby. 1998. *The relationship between Biodiversity Action Plan Priority and Broad Habitat Types, and other woodland classifications*. Peterborough: INCC.

Hofstetter, Ronald. 2000. *Universal mire lexicon*. Greifswald: IMCG.

Holden, Peter, and Tim Cleeves. 2002. *RSPB handbook of British birds*. London: A & C Black.

Interpretation manual of EU habitats EUR 27. 2007. Brussels: European Commission DG Environment.

Ilnicki, Piotr. 2002. *Torfowiska i torf*. Poznań: Wyd. Akademii Rolniczej im. A. Cieszkowskiego w Poznaniu.

Jasnowska, Janina, Mieczysław Jasnowski, and Jan Radomski. 1995. *Botanika*. Szczecin: Wydawnictwo "Brasika".

Jonsson, Lars. 1998. *Ptaki Europy*. Translated by Tadeusz Stawarczyk, Tomasz Cofta, and Jan Lontkowski. Warszawa: MUZA S.A.

Joosten, Hans, and Donal Clarke. 2002. *Wise use of mires and peatlands. A framework for decision-making*. Draft dated: 16th May 2002. International Mire Conservation Group and International Peat Society. Available at http://www.mirewiseuse.com.html-4k

—. 2001. *Identifying peatlands of international biodiversity importance*. An article available at www.imcg.net/docum/criteria.htm - 192k

—. 1998. *Mire classification for nature conservation: the dialectics of difference*. IMCG meeting and workshop on mire terminology and classification issues – Greifswald, Germany. An article available at www.imcg.net/docum/greifswa/dialect.htm - 17k

Kielak, Anna, Maria. 2004. *Zielnik Elizy Orzeszkowej. Nieznany zabytek przechowywany w zbiorach PTPN*. Poznań: Wydawnictwo Kontekst.

Kopaliński, Władysław. 1983. *Słownik wyrazów obcych i zwrotów obcojęzycznych*. Warszawa: Wiedza Powszechna.

Korsak, Włodzimierz. 1922. *Rok myśliwego*. Poznań: KN Niemierkiewicza, WKN Rzepeckiego.

Linde-Usiekniewicz, Jadwiga, ed. 2004. *Wielki słownik polsko-angielski*. Warszawa: PWN.

—. ed. 2002. *Wielki słownik angielsko-polski*. Warszawa: PWN.

Matuszewski, Gustaw, and Krzysztof Morow. 1994. *Kuraki leśne*. Warszawa: OE "Wydawnictwo Świat".

Mayr, Ernst. 1996. What is a species and what is not?. *Philosophy of Science*, vol. 63: 262-277.

Metera, Dorota, and Tomasz Pezold, eds. 2003. *Glossary of agri-environmental terms*. Warszawa: ABC Poland.

Parkyn, L., R. E. Stoneman, and H. A. P. Ingram, eds. 1997. *Conserving peatlands*. Wallingford: CAB International.

Rąkowski, Grzegorz. 2001. *Czar Polesia*. Pruszków: Oficyna Wydawnicza „Rewasz".

Ruuhijarvi, Rauno. 1983. The Finnish mire types and their regional distribution. In Mires: swamp, bog, fen and moor, vol. B., edited by AJP Gore *Ecosystems of the World, 4*. Amsterdam: Elsevier, pp. 47-67.

Sinclair, John, ed. 1995. *Collins Cobuild English dictionary*. London: HarperCollins*Publishers*.

Sjörs, Hugo. 1950. On the relation between vegetation and electrolytes in North Swedish mire waters. In *Oikos*, Vol. 2, Fasc. 2.: 241-258.

Skoczylas, Rafał. 1990. *Co w lesie piszczy?* Warszawa: Wydawnictwa Radia i Telewizji.

Smartwood. 2004. *Practical conservation through certified forestry. Forest management public summary for Regional Directorate of Forest in Radom and Łódź*. Available at www.smartwood.org.

Sokołowski, Aleksander, Anna Kliczkowska, and Mirosław Grzyb. *Określenie jednostek fitosocjologicznych wchodzących w zakres siedliskowych typów lasu*. Prace IBL seria B 32.

Sokołowski, Jan. 1992. *Ptaki Polski*. Warszawa: WSiP.

Stanisławski, Jan. 1989. *Wielki słownik polsko-angielski*. Warszawa: Wiedza Powszechna.

—. 1990. *Wielki słownik angielsko-polski*. Warszawa: Wiedza Powszechna.

Stidworthy, John. 2001. *Creepy crawlies*. Bath: Parragon.

Szujecki, Andrzej. 1998. *Entomologia leśna*. Warszawa: Wydawnictwo SGGW.

Szymański, Stanisław. 2000. *Ekologiczne podstawy hodowli lasu*. Warszawa: PWRiL.

Šomšak, Ladislav, ed. 1993. *Świat roślin, skał i minerałów*. Warszawa: MULTICO.

The management of semi-natural woodlands. 2003. Edinburgh: Forestry Commission.

The praise of honey. Honey for health, beauty and cookery. 2005. The Boyne Valley Company. Mell., Drogheda, Co. Louth. Ireland.

Tomiałojć, Ludwik, and Tadeusz Stawarczyk. 2003. *Awifauna Polski. Rozmieszczenie, liczebność i zmiany*. Wrocław: PTPP "pro Natura".

Urbańczyk, Stanisław, ed. 1953. *Słownik staropolski*. Warszawa: PAN.

Wheeler, B. D., and M. C. F. Proctor, 2000. Ecological gradients, subdivisions and terminology of north-west European mires. *Journal of Ecology* 88: 187-203.

Wołejko, Lesław, Robert Stańko, Paweł Pawlaczyk, and Andrzej Jermaczek. 2004. *Poradnik ochrony mokradeł w krajobrazie rolniczym*. Świebodzin: Wydawnictwo Klubu Przyrodników.

Zaręba, Ryszard. 1986. *Puszcze, bory i lasy Polski*. Warszawa: PWRiL.

Ziółkowska, Maria. 1983. *Gawędy o drzewach*. Warszawa: LSW.

Periodicals

Countryside. Jan/Feb. 1994

Countryside. July/August. 1993

Countryside. May/June. 1991

Internet Sources

http://animaldiversity.ummz.umich.edu/site/topics/mammal_anatomy/horns_and_antlers.html%20-%2023k

http://www.answers.com/topic/backwoods - 44k

http://www.birdforum.net/archive/index.php/t-14263.html%20 - 87k

http://www.defra.gov.uk

http://www.dragonflysoc.org.uk

http://www.forestry.gov.uk/statistics

http://www.geog.leeds.ac.uk/projects

http://www.iiasa.ac.at/Research/FOR/forest_cdrom/english/for_man_en.html - 19k

http://www.jncc.gov.uk

http://www.madametulbot.com/pix/posters/vampire2.htm_11k

http://www.nbn.org.uk/habitats/ - 11k

http://www.ruralni.gov.uk

http://www.sjp.pwn.pl

http://www.staffs-ecology.org.uk/ - 8k
http://www.staresady.republika.pl/-3k
http://www. szkolkarstwo.pl/article.php?id=492-16k
http://www.ukbap.org.uk
http://www.wiedza.servis.pl/haslo/Systematyka_gleb-38k (19 February 2006).
http://www.yourdictionary.com/backwoods - 17k

Nature terminology: primary sources analysed in the book and/or used in the database

Bousfield, Jonathan, and Mark Salter. 2005. *The rough guide to Poland.* New York. London. Delhi: Rough Guides.

Budrewicz, Olgierd. 2002. *Warmia Mazury.* Translated by Teresa Bałuk-Ulewiczowa. Olszanica: BOSZ.

Burnett-Hodgson, Francis. 2004. *Tajemniczy ogród.* Translated by Jadwiga Włodarkiewicz. Warszawa: Prószyński i S-ka.

—. 1999. *Tajemniczy ogród.* Translated by Rafał Dawidowicz. Bielsko-Biała: Wydawnictwo Klasyka. .

—. 1998. *Tajemniczy ogród.* Translated by Anna Staniewska. Wrocław: Siedmioróg.

—. (first published 1888). *The secret garden.* http://www. online-literature.com/burnett/_17k (17 March 2006).

Bzura, Waldemar, and Peter Scherbuk. 2007. *Dolina Rospudy. The Rospuda Valley.* Translated by Michał Pawica and Aleksander Ptak. Kraków: Biały Kruk sp.z.o.o.

Chylarecki, Przemysław, and Gerard Sawicki. 2003. *Ostoja Ptaków Dolina Środkowej Wisły.* Translated by Marek Czekański. Warszawa: Askon. (the title is not translated, Polish and English versions are placed on neighbouring pages).

Durrell, Gerald. 1975a. *The zoo in my luggage.* London: Penguin Books.

—. 1991. *Zoo w walizce.* Translated by Anna Wiśniewska-Walczyk. Warszawa: Wyd. "Alfa".

—. 1976. *Birds, beasts and relatives.* Harmondsworth: Penguin Books.

—. 1999. *Moje ptaki, zwierzaki i krewni.* Translated by Anna Przedpełska-Trzeciakowska. Warszawa: Prószyński i S-ka.

—. 1974. *The Whispering Land.* Harmondsworth: Penguin Books.

—. 2002. *Kraina szeptów.* Translated by Ewa Horodyska. Warszawa: Prószyński i S-ka.

—. 1973. *The Bafut beagles.* Harmondsworth: Penguin Books.

—. 2001. *Ogary z Bafut*. Translated by Ewa Horodyska. Warszawa: Prószyński i S-ka.

—. 1975b. *My family and other animals*. Harmondsworth: Penguin Books.

—. 1999. *Moja rodzina i inne zwierzęta*. Translated by Anna Przedpełska-Trzeciakowska. Warszawa: Prószyński i S-ka.

Evans, David. 1992. *A history of nature conservation in Britain*. London and New York: Routlegde.

Fowles, John. 1978. *Daniel Martin*. New York: NAL.

—. 2000. *Daniel Martin*. Translated by Jerzy Łoziński. Poznań:Zysk i S-ka.

—. 1975. *The ebony tower*. New York: New American Library.

—. 1994. *Hebanowa wieża*. Translated by Tomasz Bieroń. Poznań: Zysk i S-ka.

—. 1977. *The magus*. London: Pan Books.

—. 1982. *Mag*. Translated by Ewa Fiszer. Warszawa: PIW.

—. 1996. The French lieutenant's woman. London: Vintage.

—. 2003. *Kochanica Francuza*. Translated by Wacława Komarnicka. Warszawa: Editions Altaya Polska z.o.o, De Agostini Polska Sp. z.o.o.

—. 1985. *A maggot*. London and Sydney: Pan Books.

—. 1997. *Larwa*. Translated by Irena Doleżal-Nowicka. Poznań: Zysk i S-ka.

Fronczak, Krzysztof. 2003. *Poland's green treasurehouse*. Warszawa: CILP.

Głowaciński, Zbigniew, ed. 2001. *Polska czerwona księga zwierząt. Polish red data book of animals*. Translated by Małgorzata Makomaska-Juchiewicz . Warszawa: PWRiL.

Grootjans, Ab, and Lesław Wołejko, eds. 2007. *Conservation of wetlands in Polish agricultural landscapes. Ochrona mokradeł w rolniczych krajobrazach Polski*. Szczecin: Oficyna In Plus.

Hryniewiecki, Tomasz, and Grzegorz Okołów. 2004. *Echo mazowieckiej kniei. Echo of Mazovian forest*. Translated by Barbara Diehl. Warszawa: Muza SA.

Iwaszkiewicz, Jarosław. 2002. *The Birch Grove and other stories*. Translated by Antonia Lloyd-Jones. Budapest London New York: CEU Press.

—. 1987. *Brzezina i inne opowiadania ekranizowane*. Warszawa: Czytelnik.

—. 1969. *Młyn nad Utratą*. In: *Opowiadania zebrane*. Warszawa: Czytelnik, pp. 249-341.

Kamieniarz, Robert. 2002. *Cietrzew*. Swiebodzin: Wydawnictwo Klubu Przyrodników.(it contains an English summary, translator unknown).

Karolczuk-Kędzierska, Monika. 2007. *Poland. Treasures of nature.* Translated by Władysław Chłopicki. Kraków: Wyd. Kluszczyński.

Kaźmierczakowa, Róża, and Kazimierz Zarzycki, eds. 2001. *Polska czerwona księga roślin. Polish red data book of plants.* Translated by Małgorzata Makomaska-Juchiewicz. Kraków: Wydawnictwo Instytutu Botaniki PAN.

Kłosowscy, Grzegorz, Stanisław and Tomasz. 2005. *Polska portret przyrody. Poland nature portraits.* Translated by Katarzyna Tomkiewicz. Warszawa: MULTICO.

—. 2003. *Biebrza. Ptasi raj. The Biebrza. A bird paradise.* Translated by Katarzyna Tomkiewicz. Warszawa: MULTICO.

—. 1999. *Polska. Cuda natury. Poland. Natural wonders.* Translated by Katarzyna Tomkiewicz. Warszawa: MULTICO.

Kobyliński, Szymon. 1984. *Jej Królewska Mość Wisła. Her Majesty the Vistula.* Translated by Doris Ronowicz. Warszawa: Interpress.

Kossak, Simona. 2001. *Saga Puszczy Białowieskiej.* Warszawa: MUZA S.A.

Kossak, Simona. 2001. *The Białowieża Forest saga.* Translated by Elżbieta Kowalewska. Warszawa: MUZA S.A.

Kot, Henryk, and Andrzej Dombrowski, eds. 2001. *Strategia ochrony fauny na Nizinie Mazowieckiej. Strategy for conservation of fauna in the Mazovian Lowland.* Translated by Magdalena Remisiewicz. Siedlce: MTOF.

Krogulec, Jarosław, ed. 1998. *Ptaki łąk i mokradeł Polski.* Warszawa: Fundacja IUCN Poland (it contains an English summary, translator unknown).

Kucharski, Leszek, and Włodzimierz Pisarek. undated. *Polany Puszczy Bolimowskiej. The glades of Bolimów Woodland.* Skierniewice: Dyrekcja Bolimowskiego Parku Krajobrazowego (translator unknown).

Lewandowska, Izabela, and Dariusz Lewandowski, eds. 1994. *Mazowiecki i Chojnowski Park Krajobrazowy. Mazovian and Chojnów Landscape Park.* Warszawa: Zarząd MiChPK (translator unknown).

Lewandowska, Izabela, and Dariusz Lewandowski, eds. 1999. *Mazowieckie Parki Krajobrazowe. Mazovian Landscape Parks.* Warszawa: Zarząd MiChPK (translator unknown).

Łapińscy, Włodzimierz, and Rafał. 2007. *Rospuda. Bagno i rzeka. The Rospuda. Wildlife of valley.* Translated by Magda Podlaska. Bryzgiel: Wyd. W. Łapiński.

Mickiewicz, Adam. 1996. *Pan Tadeusz.* Warszawa: ELIPSA.

—. 1964. *Pan Tadeusz*. Translated by Kenneth MacKenzie. London: The Polish Cultural Foundation.

Miłosz, Czesław. [1955] 1989. *Dolina Issy*. Kraków: Wydawnictwo Literackie.

—. 1981. *The Issa Valley*. Translated by Louis Iribarne. London: Sigdwick & Jackson.

Narodowa Fundacja Ochrony Środowiska. Biebrzański Park Narodowy. 1996. *Biebrzański Park Narodowy. Opis przyrodniczy. Biebrza National Park. Description of natural conditions*. Translated by Jerzy Rolnik and James R. A. Richards. Białystok: OPGK S-ka z o.o.

Nowiński, Krzysztof. 2005. *Polska. Opowieść o ludziach, zabytkach i przyrodzie. Poland. People, nature and historical treasures*. Translated by Dorota Bartnik. Warszawa: Sport i Turystyka. Muza S.A.

Nycz, Beata, ed. 2004. *Polska. Klejnoty przyrody. Poland. Jewels of nature*. Translated by Michael Jacobs. Bielsko Biała: Buffi.

Okołów, Czesław, and Grzegorz Okołów. 2005. *Białowieża National Park*. Translated by Wojciech Kasprzak. Warszawa: MULTICO.

—. 2001. *Białowieski Park Narodowy*. Warszawa: MULTICO.

Ośrodek Edukacyjno-Muzealny RPN. 2000. *Roztoczański Park Narodowy. Mapa turystyczno-przyrodnicza* 1: 50000. Lublin: Kartpol.

Page, Robin, ed. 2002. *The feather and the furrow. The bird photographs of Chris Knights*. Cambridge: Bird's Farm Books.

Pawlaczyk, Paweł, Jacek Herbich, Jan Holeksa, Jerzy Szwagrzyk, and Krzysztof Świerkosz. 2003. *Rozpoznawanie siedlisk przyrodniczych na podstawie danych opisu taksacyjnego lasu*. Manuscript. Available at http:// www.kp.org.pl/n2k.

Pawlaczyk, Paweł, and Wojciech Mróz. 2003. *Natura 2000 a gospodarka leśna*. In *Natura 2000 w lasach Polski. Skrypt dla każdego,* edited by Małgorzata Buszko-Briggs. Manuscript. Białowieża: COWI, pp. 56-163.

Polakowski, Benon, Antoni Jutrzenka-Trzebiatowski, and Czesław Hołdyński. 1997. *Roślinność rzeczywista Mazurskiego Parku Krajobrazowego. Actual vegetation of the Masurian Lanscape Park. Mapa przyrodniczo-turystyczna 1:30000*. Olsztyn: Wyd. Art. Olsztyn.

Przystalski, Andrzej, and Tomasz Załuski. 1997. *Lasy łąckie. The Łąck Forests*. Translated by Anita Lewicka. Kowal: Zarząd Parków Krajobrazowych Brudzeńskiego i Gostynińsko-Włocławskiego.

Pudlis, Eugeniusz. 2003. *13 wrót do leśnego królestwa. 13 gates to the forest kingdom*. Translated by Richard Biały. Warszawa: CILP.

Rodziewiczówna, Maria. 1998. *Lato leśnych ludzi*. Warszawa: Prószyński i S-ka.

Różycki, Andrzej, and Arkadiusz Iwaniuk. 2001. *Poleski Park Narodowy. Poleski National Park*. Translated by Magdalena Butrym-Karaś. Urszulin: Brokart Mat.

Rysiewicz, Maciej. 2000. *Ule i pasieki w Polsce. Hives and apiaries in Poland*. Translated by Marta Białecka and Tom Allan. Stróże: Wyd. "Sądecki Bartnik".

Sawicki, Gerard. 2003a. *Ostoja Ptaków Ujście Wisły*. Translated by Marek Czekański. Warszawa: Askon (the title is not translated, Polish and English versions are placed on opposite pages).

—. 2003b. *Lasy Iławskie*. Translated by Andrzej Kowalczewski. Warszawa: Askon (the title is not translated, Polish and English versions are placed on opposite pages).

Sidło, Paweł Olaf, Bogumiła Błaszkowska, and Przemysław Chylarecki, eds. 2004. *Ostoje ptaków o randze europejskiej w Polsce. Important Bird Areas of European Union importance in Poland*. Translated by Paweł O. Sidło, Magda Remisiewicz, Tomasz Kundera, Zoltan Waliczky. Warszawa: OTOP.

Промо: Sobociński, Wojciech. 2004. *Kraina żubra. The land of the bison*. Translated by Andrzej Piotr Zakrzewski. Ciemięrzowice: Wyd. "globalna wioska".

The Monarch of the Glen. 2000-2005. TV. BBC1. Ecosse Films.

Tumiłowicz, Jerzy, and Piotr Banaszczak. 2000. *Arboretum w Rogowie. The Rogów arboretum*. Warszawa: MULTICO.

Wiatr, Artur, and Włodzimierz Wróblewski. 2005. *Biebrza National Park*. Translated by Wojciech Kasprzak. Warszawa: MULTICO.

—. 2000. *Biebrzański Park Narodowy*. Warszawa: MULTICO.

Wildlife Aid. Animal Planet. Shown 22nd Nov 2007.

Witczuk, Julia. 2001. *Siberia – In the Baikal Land of the Buryats*. Translated by Katarzyna Tomkiewicz. Warszawa: MULTICO.

—. 2001. *Syberia – W nadbajkalskiej krainie Buriatów*. Warszawa: MULTICO.

Załącznik do Zarządzenia nr 43 Dyrektora Generalnego Lasów Państwowych z 18.04.2003 r.

Zajączkowski, Grzegorz. 1999. *Lasy Państwowe. Raport roczny. The State Forests. Report on activities for 1999*. Translated by James Richards. Warszawa: CILP.

Zawadzcy, Dorota, and Jerzy. 2003. *Głuszec*. Świebodzin: Wydawnictwo Klubu Przyrodników (it contains an English summary, translator unknown).

Internet Sources

http://www.lkp.org.pl/poradniki/lasy/fotoprzewodnik_lesne.pdf
http://www.lkp.org.pl/poradniki/lasy/fotoprzewodnik_nielesne.pdf

INDEX